Innovation Without Patents

Innovation Without Patents

Harnessing the Creative Spirit in a Diverse World

Edited by

Uma Suthersanen

Reader in Intellectual Property Law and Policy, Queen Mary, University of London, UK

Graham Dutfield

Herchel Smith Senior Research Fellow in Intellectual Property Law, Queen Mary, University of London, UK

Kit Boey Chow

Singapore IP Academy, Singapore

Based on a Report Commissioned by the Singapore Property Academy

Edward Elgar

Cheltenham, UK • Northampton, MA, USA

Published by
Edward Elgar Publishing Limited
Glensanda House
Montpellier Parade
Cheltenham
Glos GL50 1UA
UK

Edward Elgar Publishing, Inc.
William Pratt House
9 Dewey Court
Northampton
Massachusetts 01060
USA

A catalogue record for this book
is available from the the British Library

Library of Congress Cataloging in Publication Data
Innovation without patents : harnessing the creative spirit in a diverse world / edited by Uma Suthersanen, Graham Dutfield, and Kit Boey Chow.
p. cm.
Includes bibliographical references and index.
1. Patent laws and legislation. 2. Technological innovations—Law and legislation. 3. Intellectual property. 4. Inventions. I. Suthersanen, Uma. II. Dutfield, Graham. III. Chow, Kit Boey, 1944-
K1505.I56 2007
346.04'86—dc22

2006014304

ISBN: 978 1 84542 959 1

Typeset by Manton Typesetters, Louth, Lincolnshire, UK
Printed and bound in Great Britain by MPG Books Ltd, Bodmin, Cornwall

Contents

Contributors vii
Acronyms viii
Foreword ix
Acknowledgements x

PART ONE INNOVATION, DEVELOPMENT AND INTELLECTUAL PROPERTY

1 Innovation and development 3
Graham Dutfield and Uma Suthersanen
2 Innovation and the law of intellectual property 13
Uma Suthersanen and Graham Dutfield
3 Utility models and other alternatives to patents 18
Uma Suthersanen and Graham Dutfield
4 Policy considerations for governments 64
Uma Suthersanen and Graham Dutfield

PART TWO HARNESSING MINOR INNOVATION: NATIONAL STUDIES

5 Singapore 73
Kit Boey Chow, Kah Mun Leo, Susanna Leong
6 Australia 119
Andrew Christie and Sarah Moritz
7 Japan and South Korea 142
Graham Dutfield and Uma Suthersanen
8 China and Taiwan 152
Kit Boey Chow, Kah Mun Leo, Susanna Leong with Jerry Hsiao
9 The ASEAN states 169
Uma Suthersanen with Lim Heng Gee
10 Utility models in Latin America 188
Manuel Marquez

PART THREE DIVERSE PATHS TO AN INNOVATIVE FUTURE

11 Conclusions and recommendations 195
Uma Suthersanen and Graham Dutfield

Index 201

Contributors

Graham Dutfield, Queen Mary, University of London

Uma Suthersanen, Queen Mary, University of London

Kit Boey Chow, Singapore IP Academy

Kah Mun Leo, Singapore IP Academy

Susanna Leong, National University of Singapore

Lim Heng Gee, University Teknologi Mara, Malaysia

Andrew Christie, Intellectual Property Research Institute of Australia

Sarah Moritz, Intellectual Property Research Institute of Australia

Jerry Hsiao, Queen Mary, University of London

Manuel Marquez, L.L. & M.M. Consultants, Mexico City

Acronyms

ARIPO	African Regional Industrial Property Organization
ASEAN	Association of Southeast Asian Nations
CDR	Community Design Right
EU	European Union
GDP	gross domestic product
IP	intellectual property
IPC	International Patent Classification
OAPI	African Intellectual Property Organization
OECD	Organisation for Economic Co-operation and Development
PCT	Patent Cooperation Treaty
R&D	research and development
SIPO	State Intellectual Property Office (China)
SME	small and medium enterprise
TRIPS	Agreement on Trade-related Aspects of Intellectual Property Rights
UM	utility model
USPTO	United States Patent and Trademark Office
WIPO	World Intellectual Property Organization
WTO	World Trade Organization

Foreword

'Innovation patents' in Australia, 'petty patents' in Indonesia and Thailand, 'utility models' in Cambodia, China and the Philippines, 'new utility models' in Taiwan, 'utility innovations' in Malaysia, 'utility solutions' in Vietnam. And that is only the patent-like protection given to minor innovations; in addition, around the globe there are widely differing systems providing protection against 'unfair' copying.

When there are so many different ways of describing what is supposed to be the same intellectual property righty, it indicates surely that this is a difficult and controversial area of the law. So, in July 2004 Professor Gerald Dworkin, my predecessor as Director of the IP Academy Singapore, decided that this would be an interesting topic for the newly-created Academy to commission an international and comparative study.

In turn, to lead the project Professor Dworkin looked to his former colleague at Queen Mary College in London, Dr Uma Suthersanen, who is not only a native Singaporean but also a world-renowned expert of design laws.

Dr Suthersanen has pulled together an outstanding team of fellow experts from her College (Graham Dutfield), from Singapore (Kit Boey Chow, Kah Mun Leo and Susanna Leong), from Australia (Andrew Christie and Sarah Moritz), from Taiwan (Jerry Hsiao) from Malaysia (Lim Heng Gee) and Mexico (Manuel Marquez)

The resulting book is a fascinating comparative insight into the different ways that countries have tried to tackle the issues both of minor inventions falling below the standard required by patent law and of 'free riding' on product design and appearance.

At the end of it all, Drs Suthersanen and Dutfield offer up for consideration by developing countries three options: the status quo approach, the accretion approach and the emulation approach. But, as with any good book, I will not spoil the ending for you.

Professor David Llewelyn
Director, IP Academy Singapore

Acknowledgements

We thank Gerald Dworkin, David Llewelyn, Ng Wee Loon, Charlotte Tan and the entire Singapore IP Academy team for their invaluable support and patience during the lifetime of this research project from which much of the text of this book resulted. The project also benefited from the financial support in relation to UM protection in Europe from the International Centre for Trade and Sustainable Development, and many thanks to Christoph Spennemann of UNCTAD who offered invaluable suggestions and comments especially in relation to the sections on Europe and Germany.

Rod Crawford and Sean Applegate from IP Australia, Sophie Waller, Sally Pryor, R. Mitch Casselman, Paul Jensen and Peter Summers are thanked for their assistance with the Australian chapter. Thanks also to Lim Kok Seng for the invaluable survey support in relation to the Singapore chapter. Finally, our thanks to Yasmine Meajun and Cristina Garrigues for their able assistance in the initial gathering of data for the EU and Germany for the report.

PART ONE

Innovation, development and intellectual property

1. Innovation and development

Graham Dutfield and Uma Suthersanen

1.1 DEVELOPMENT AND DIVERSITY

'Development', a word borrowed from biology, is a term whose meaning is contested by social scientists and international development experts and organisations. Nowadays, it is common to speak of 'economic development', which focuses on a country's measurable economic performance relative to other countries; of 'human development', which supplements economic development by incorporating social welfare considerations; and of 'sustainable development', which takes into account the environment as well.

Conventionally, the extent of a country's development is quantified by using certain indicators of income and output, such as gross national product (GNP). This is, of course, *economic* development. At its crudest, the economic progress of different countries is compared by making country league tables, with the richest nations according to GNP per capita at the top and the poorest with the lowest figures propping up the table at the bottom. The World Bank's annual World Development Reports rank countries in this way (although the reports provide various other indicators of development as well).[1]

More commonly, though, we tend to talk of developed countries and developing countries as if there are no other kinds of country. Alternatively, but similarly, following the 1980 report of the Independent Commission on International Development Issues chaired by former West German Chancellor Willi Brandt, the developed world is 'the North' and the developing world is 'the South'.[2] In 1971, though, the United Nations (UN) carved out a sub-category of the latter grouping, the 'least developed countries' (LDCs). At present there are around 50 of these countries, which are considered to be the poorest of the poor. LDCs receive preferential treatment in terms of development finance, trade and technical assistance. More than half of these countries are in Africa.

1 For synopses on present and past World Development Reports, see http://econ.worldbank.org.

2 Brandt, W. (1980), *North–South: A Programme for Survival*, London: Pan.

Even with the addition of the LDCs, this is rather simplistic and fails utterly to reflect the sheer diversity of the world, which is extreme even if we confine our analysis to those countries that are poor compared to Western Europe, North America, parts of East Asia and Australasia. Indeed, at the opposite end of the developing world spectrum from the LDCs, and far far away in terms of income levels and living standards, are the vastly different newly industrialising economies, those like Singapore, Taiwan, South Korea and Hong Kong, which have attained a level of economic development that is very close to that of developed countries and, rather illogically, exceeds some of them. There are also the transition economies consisting of the former communist countries that are converting to market economies but are not wealthy enough yet to be classed as developed countries.

GNP per capita paints a very incomplete, not to say misleading, picture of the actual conditions and capabilities of individual people in a country. As the Indian economist and Nobel laureate Amartya Sen explains, 'there is every evidence that even with relatively low income, a country that guarantees health care and education to all can actually achieve remarkable results in terms of the length and quality of life of the entire population'.[3] In order to make the concept of development more people oriented, the UN Development Programme, influenced by the work of Sen, adopted the term 'human development'. The UNDP human development index, which was devised by Pakistani economist Mahbub ul Haq, and is updated every year in the organisation's Human Development Reports, ranks countries by their performance in terms of life expectancy, literacy, education and gross domestic product.[4]

The concept of 'sustainable development' has its origins in the 1980 World Conservation Strategy of the International Union for the Conservation of Nature and Natural Resources.[5] Sustainable development was seen as a normative process in which conservation and development are combined to improve human welfare while practising what the strategy refers to as 'living resource conservation'.

According to the 1987 World Commission on Environment and Development, chaired by former Norwegian prime minister Gro Harlem Brundtland, sustainable development is 'development that meets the needs of the present without

---

[3] Sen, A. (1999), *Development as Freedom*, Oxford: Oxford University Press, p. 144.

[4] To download present and past Reports, visit the UNDP–HDRO website: http://hdr.undp.org/.

[5] IUCN, UNEP and WWF (1980), *World Conservation Strategy: Living Resource Conservation for Sustainable Development*, Gland: IUCN.

compromising the ability of future generations to meet their own needs'.[6] Sustainable development embodies two key notions:

- the concept of 'needs', in particular the essential needs of the world's poor, to which overriding priority should be given
- the idea of limitations imposed by the state of technology and social organisation on the environment's ability to meet present and future needs.

At the 1992 UN Conference on Environment and Development (the 'Earth Summit') in Rio de Janeiro, the international community embraced the concept of sustainable development when it adopted Agenda 21, a programme to promote sustainable and environmentally sound development from then and into the twenty-first century, and the Rio Declaration on Environment and Development.

In 2000, the UN Millennium Summit was held at which UN member states agreed on a set of goals and targets for achieving development. The eight goals are now known as the Millennium Development Goals (MDGs), and are as follows:

1. Eradicate extreme poverty and hunger
2. Achieve universal primary education
3. Promote gender equality and empower women
4. Reduce child mortality
5. Improve maternal health
6. Combat HIV/AIDS, malaria and other diseases
7. Ensure environmental sustainability
8. Develop a global partnership for development.

The version of development that best captures the intent of the MDGs is sustainable development. This is as good a reason as any to consider that 'development' in this book should be understood as sustainable development.

1.2 INNOVATION AND ITS ROLE IN DEVELOPMENT

Innovation is not necessarily lacking in developing countries. However, harnessing innovation to generate wealth is a huge challenge for many of them. Of

---

[6] World Commission on Environment and Development (1987) *Our Common Future*, Oxford: Oxford University Press, p. 43.

course, some developing countries are already quite advanced in science, technology and industrial development and are well placed to exploit the creativity of their populaces. The majority of countries, though, still have a long way to go indeed.

Enhancing the scientific and technological bases of developing countries requires appropriate regulatory and legal frameworks providing rewards and incentives for innovation and investment. For them to transform their comparative advantage from producing low-value commodities to high-value goods and services while increasing employment possibilities for the poor, developing countries need to identify technological fields or industrial or market sectors where they may be able to compete internationally, and to acquire and/or channel the necessary investments to develop and market high value products. While science-based research-intensive industries and technologies like information and communications technologies, pharmaceuticals and the new biotechnologies are extremely important, competitive high-value-added products can be developed without cutting-edge scientific knowledge and equipment. High-value products may succeed in the market based on knowledge acquired from a range of local sources including traditional communities, which often have to be creative just to survive. Products will command high prices in international markets only if they are knowledge-intensive, but this does not by definition require them to be science-based, high-tech or R&D-intensive. In fact, various kinds of knowledge must be acquired and used for *any* product to succeed in the increasingly competitive global economy. These need not be science and technology-related at all. According to Mytelka and Tesfachew,[7] these kinds of knowledge include:

- product design
- process engineering
- quality control
- management and maintenance routines
- knowledge about markets and investment opportunities
- skills and capabilities needed to undertake changes in products and processes, create networks and sustain partnering activity.

This much broader conception of 'knowledge' comes close to Schumpeter's definition of 'innovation' (i.e. 'carrying out new combinations'),[8] which comprises:

---

[7] Mytelka, L.K. and T. Tesfachew (1998), *The Role of Policy in Promoting Enterprise Learning During Early Industrialization: Lessons for African Countries*, Geneva: United Nations Conference on Trade and Development, p. 2.

[8] Schumpeter, J.A. (1983 [1934]), *The Theory of Economic Development: An Inquiry into Profits, Capital, Credit, Interest, and the Business Cycle*, New Brunswick, NJ: Transaction Publishers, p. 66.

1. 'the introduction of a new good ...
2. the introduction of a new method of production ... *which need by no means be founded upon a discovery scientifically new* ...
3. the opening of a new market ...
4. the conquest of a new source of supply of raw materials ...
5. the carrying out of the new organization of any industry ...' [emphasis added].

Innovation as understood in this broad sense is essential for countries' shift from producing low-value raw materials to high-value-added manufactured goods and commodities.

Innovation can be categorised as 'discrete' or 'cumulative'. Discrete innovation is independent of previous innovation.[9] In reality however, creation does not occur in a vacuum. A vast majority of scientific and cultural creations, if not all, are built on pre-existing creations and discoveries and do not represent giant leaps beyond what we already know.[10] This type of innovation can be said to be cumulative in that it is based on multiple small steps or increments. New 'things' are produced by combining existing elements to produce new combinations or variations. The fact that innovation builds on innovation in a cumulative manner reflects the two economic characteristics of information: non-exhaustibility and non-excludability. Mandeville notes that by accepting the reality that innovation is cumulative, one can extrapolate two points: first, although information cannot be exhausted, new or value-added information will be added to the existing stock of knowledge; second, since excludability is difficult and indeed dangerous for innovation, it may be good for both scientific and cultural industries to share innovation as widely as possible.[11] This must be so or else the economic and social costs of constantly re-creating knowledge would be incalculable. Indeed, efficient creation of new works and inventions requires access to and use of old works, especially if the by-product of such a practice is that the follow-on work is superior to the pre-existing work.[12]

---

[9] Merges, R. and R. Nelson (1990), *On the Complex Economics of Patent Scope*, 90 Colum. L. Rev. 839, 880–1, citing examples in the toy and consumer product industry.

[10] Scotchmer, S. (1999), 'Cumulative Innovation in Theory and Practice', University of Berkeley working paper series, February, 1; Cohen, Julie E. and Mark A. Lemley (2001), *Patent Scope and Innovation in the Software Industry*, 89 Cal. L. Rev. 1, 5 (the software industry 'is characterized by rapid sequential innovation'); also Merges, R. and R. Nelson, *On the Complex Economics of Patent Scope*, *op. cit.*, 880–1.

[11] Mandeville, T. (1998), 'An information economics perspective on innovation', *Journal of Social Economics*, **25**, 360.

[12] Merges, R. and R. Nelson (1991), *On the Complex Economics of Patent Scope*, *op. cit.*, 908; Scotchmer, S. (1991), 'Standing on the shoulders of giants: cumulative research

Furthermore, innovation does not occur within a natural or legal individual but tends to arise from social interaction which involves both creators and other actors whereby various sources of information are assembled to create a technology or work which builds on and enhances previously existing forms.[13] Schumpeter points to a similar phenomenon when he discusses economic leadership. Ideas and creations, he says, 'are always present, abundantly accumulated by all sorts of people. Often they are also generally known and being discussed by scientific or literary writers'. However, society requires an 'economic leader' to amass all these things and to present it to society. In underlining the importance of other economic actors in bringing forth an invention, Schumpeter offers the secondary, and much narrower, basis of innovation as something which an economic leader must do in order to render an invention practical and acceptable to societal use.[14] A correlated view is that most innovations, as opposed to groundbreaking and dramatic inventions, are routine and primarily devoted to product improvement or enhanced user-friendliness or searches for new uses for those products. However, there is a synergy between the two types of inventive activity, and both types of activities are vital in enabling societal advance and growth.[15] Many economists, by focusing on Schumpeter's narrow usage of innovation, see innovation as something which introduces new technology on the market.[16] This tends to view innovation as something inextricably linked to patent systems. However, as Scherer points out, innovation does occur in cultural spheres such as music performance and composition, film making and book writing.[17] Innovation must be understood in the broader Schumpeterian sense as set out in the five factors just listed: the usage of information in the creation of scientific and cultural goods.

In sum, innovation is a complex process intertwined with factors such as the strength of the knowledge base, institutional arrangements, qualifications of the labour force, openness of the economy and an overall ability to take on board improvements achieved in other countries or sectors.[18]

---

and the patent law', *Journal of Economic Perspectives*, **29**, 30–1; Lemley, M. (1997), 'The economics of improvement in intellectual property law', 75 Texas L. Rev. 989, 997.

13 Mandeville, *op. cit.*, 360.

14 Schumpeter, *op. cit.*, 88–9.

15 Baumol, W. (2002), *The Free-Market Innovation Machine*, Princeton, NJ: Princeton University Press, p. 22; Rosenberg, N. (1976), *Perspectives on Technology*, Cambridge: Cambridge University Press, p. 66.

16 Note Merges' very narrow perspective: 'An innovation is the "debugged" and functional version of the invention: the version first offered for sale', Merges, R.P. *Commercial Success and Patent Standards*, *op. cit.*, 807, 845 *cf* 851; Nelson, R. and S. Winter (1982), *An Evolutionary Theory of Economic Change*, Cambridge, MA: Belknap Press, p. 263.

17 Scherer, F.M. 'The Innovation Lottery', *op. cit.*, 12.

18 Dutfield, G. and U. Suthersanen (2004), 'The innovation dilemma: intellectual property and the historical legacy of cumulative creativity', *Intellectual Property Quar-*

Innovation connotes newness but it is possible to argue that an innovation for one company or national economy may not necessarily be innovative to another. Ernst et al.[19] make this point when they define 'innovation' as:

> the process by which firms master and implement the design and production of goods and services that are new to them, *irrespective of whether or not they are new to their competitors – domestic or foreign* [emphasis added].

Although this definition has the possible weakness of blurring the distinction between innovation and imitation, it does at least make clear that promoting innovation in developing countries also means facilitating the acquisition, dissemination and (where necessary) adaptation of knowledge and technologies from elsewhere.

While it is the private sector that will be most involved in external trade, governments have a vital role to play in capacity building and in creating a conducive institutional and regulatory environment to promote innovation from basic research to commercialisation. According to UNCTAD:[20]

> Today more than ever before, it is recognized that the production and distribution of generic knowledge, be it in the form of basic science or basic general training, is a responsibility of governments. The incentives for the private sector are not strong enough to guarantee an adequate level of investments in these.

It is widely accepted among policymakers that governments, the private sector and other stakeholders need to cooperate in order to generate, attract, diffuse and exploit technological innovation for the benefit of the national economy. It falls beyond the scope of this book to consider in detail all the elements of the institutional/regulatory environment or what in this context might be referred to as the 'national system of innovation', except for one: intellectual property rights. Use of the term 'national system of innovation' (NSI) seems incongruous in light of the increasing phenomena of international business networks and innovation chains in science and technology-based industrial sectors.[21] However,

---

terly, **4** (379); and Lee, Y. and M. Langley (2004), 'Invention and innovation', *CIPA Journal*, (August), 464.

19 Ernst, D., L.K. Mytelka and T. Ganiatsos (1998), 'Technological capabilities in the context of export-led growth: a conceptual framework', in D. Ernst, T. Ganiatsos and L.K. Mytelka (eds), *Technological Dynamism and Export Success in Asia*, London: Routledge, pp. 5–45 in Mytelka and Tesfachew 1998:1–2.

20 United Nations Conference on Trade and Development (1997), 'A Project Proposal for STIP Country Reviews', Geneva: UNCTAD, p. 5.

21 Lundvall, B.-A. (1992), 'Introduction', in B.-A. Lundvall (ed.) *National Systems of Innovation: Towards a Theory of Innovation and Interactive Learning*, London and New York: Pinter, pp. 1–19, at 3–4; Swyngedouw, E. (1999), 'Territories of innovation

NSIs are understood as being open systems with the possible exception of North Korea's (though it is unclear whether very much innovation takes place in that country anyway).

1.3 LEARNING HOW TO INNOVATE: IS COPYING NECESSARY?

A very important question arises, which is that of whether learning to innovate *requires* freedom to imitate, and if so, how much freedom should be permitted. A related question is, can imitation itself be creative? Dealing with the former question first, there is ample historical evidence to suggest that learning to innovate does indeed require such freedom. Over the years, Royal Philips Electronics has been responsible for an impressive series of breakthrough inventions, such as compact audio cassettes and compact discs. What is less well known is that the company was set up in 1891 to commercially exploit somebody else's invention: Thomas Edison's and Joseph Swan's carbon filament lamp. Commercial success generated considerable revenues that enabled the firm to produce its own inventions and eventually become one of the world's most innovative corporations. How was Philips able to get such a good head start? From 1869 until 1912, Holland had no patent law. This meant that local entrepreneurs could copy foreign inventions and put them to work for their own profit, at least as long as they could figure out how they worked.

The well-known Swedish mobile phone company, Ericsson, was formed in 1876, the same year as Alexander Graham Bell made his first phone call. Sent some of these new devices to repair, the company worked out how to make them, and by 1878 was selling its own phones to the Swedish public. Bell had neglected to file patents on his invention in Sweden, and the rest is business history.

In 1960, Texas Instruments filed a patent in Japan on the integrated circuit, arguably one of the most important inventions of the second half of the twentieth century. The Japan Patent Office allowed itself 29 years to grant the patent. By that time Japanese companies, free to read the patent specification 18 months after filing, acquired the technology, improved upon it and controlled 80 percent of the United States market for computer semiconductors.[22]

– innovation as a collective process and the globalisation of competition', in H. Lawton-Smith (ed.) *Technology Transfer and Industrial Change in Europe: The Case of the Electronic Component and the Flow Measuring Industries in the UK, France and Belgium*, London: Macmillan, pp. 17–33.

[22] Johnson, C. (1995), *Japan: Who Governs? The Rise of the Developmental State*, New York and London: W.W. Norton & Co., pp. 74–5.

Richard Arkwright's cotton spinning machine, patented in England a few years before America's independence, was copied by entrepreneurs in the latter country who did not have to worry about patents since there was no patent law until 1790. Intriguingly, the machine's obvious lack of novelty by 1791 did not stop a businessman from receiving a United States patent for it.[23] Once again, in a world of highly varied national patent laws – or no laws at all in some cases – one country's invention was also another country's economic gain.

Setting to one side the rights and wrongs of such 'borrowings', and one should add here that many other examples could be given, the point is that such behaviour broke no international rules of the day. Furthermore, freedom to use such technologies was often beneficial not only to the imitator companies but also to the national economies in which they were based. Indeed, none of the recipient countries remained copiers for long; eventually they became among the world's most technologically advanced.

If we turn to the second question, free-riding and imitation can be creative in themselves, or can at least lead to creative activity. Indeed, they may even be essential for an innovative environment, albeit within reasonable limits. According to Kim and Nelson, 'imitation ranges from illegal duplicates of popular products to truly creative new products that are merely inspired by a pioneering brand'.[24] Distinct imitations may include 'knockoffs or clones, design copies, creative adaptations, technological leapfrogging, and adaptation to another industry'.[25] One should not take this argument too far, though. Wholesale copying of CDs and misappropriation of trade marks provides no scope for learning at all. Moreover, if it is too easy to profit from uncreative imitation or substitutive copying, there is unlikely to be much incentive to innovate.

This discussion raises an important policy issue, which is one of the themes of the book. This issue concerns the extent to which innovations should *not* be protected as intellectual property, or if they are, how far the rights should be limited so that local producers can learn how to innovate. A related issue, which we will consider in the next chapter and explore throughout this book, is how far should legal protection extend to inventions that may not quite, or may only just, meet the conventional standards of inventiveness required by the patent system. Should we just set them free? Should we lower patentability thresholds? Or ought we to offer alternative intellectual property regimes to patents for these

---

23 Wallace, A.F.C. and D.J. Jeremy (1977), 'William Pollard and the Arkwright patents', *William and Mary Quarterly*, **34**, 404–25.

24 Kim, L. and R.R. Nelson (2000), 'Introduction', in L. Kim and R.R. Nelson (eds), *Technology, Learning, and Innovation: Experiences of Newly Industrializing Economies*, Cambridge: Cambridge University Press.

25 Kim and Nelson *op. cit.*, citing S. Schnaar (1994), *Managing Imitation Strategy: How Later Entrants Seize Markets from Pioneers*, New York: Free Press.

inventions? One obvious alternative is utility models, which are a kind of low cost, short term patent system for inventions that lack the inventiveness required under most conventional patent systems. One of our contentions is that such questions are relevant to all countries; however, different countries are bound to have varying answers to these questions, in part due to their disparate levels of development. Another is that coming up with sound evidence-based answers is a far from easy task.

It is our hope that this book will provide some useful evidence.

2. Innovation and the law of intellectual property

Uma Suthersanen and Graham Dutfield

2.1 THE PATENT SYSTEM

The origins of patents for inventions lie in the ancient European state privileges which granted an exclusive right with the aim of encouraging domestic innovation and exploitation of technology – indeed, 'inventive activity' was not a necessary requirement as the value lay in the dissemination of the teachings inherent in the patented technology.[1] Furthermore, the prevailing mercantilist ethos of the time accepted the principle that a system of exclusive privileges would nurture innovative activity which would, in turn, promote the economic well-being of the country. The mercantilist regarded the state as the appropriate instrument for promoting the well-being of his country; in his view the country was regarded as a unit with national interests, irrespective of the interest of particular sections of individuals. Accordingly, the state harnessed and controlled resources, skills and products for the purposes and profit of the state. Patent privileges were merely one species in the genus of privileges, charters, franchises, licences and regulations issued by the Crown or by local governments within the mercantilist framework.[2]

By the end of the eighteenth century, the general consensus, encouraged by Adam Smith and Jeremy Bentham,[3] was that the existence of the patent regime

---

1 Early examples of codified patent customs are the 1474 Venetian Decree (the decree broadly granted exclusive rights to an inventor of a machine or process, including the right to request damages and the destruction of infringing devices); and the 1623 British Statute of Monopolies 1623, 21 Jac. I c.3 (whereby an inventor of any 'manner of new manufacture' would be accorded a 14 year term of protection. The notion of inventor did not merely encompass the creator of the invention, but any person who imported the invention from abroad). See Penrose, E. (1951), *The Economics of the International Patent System*, Baltimore, MD: The Johns Hopkins Press, pp. 1–12, 88–90.

2 Machlup, F. and E. Penrose (1950), 'The patent controversy in the 19th century', *Journal of Economic History*, p. 1.

3 Adam Smith, a firm supporter of granting a temporary monopoly to inventors, viewed it as a justifiable exception to free trade principles, Smith, A. (1776), *An Inquiry*

was justified on the basis of the reward theory. J.S. Mill summarised this view:

> The condemnation of monopolies ought not to extend to patents, by which the originator of an improved process is allowed to enjoy, for a limited period, the exclusive privilege of using his own improvement. This is not making the commodity dear for his benefit, but merely postponing a part of the increased cheapness which the public owe to the inventor, in order to compensate and reward him for the service. That he ought to be both compensated and rewarded for it, will not be denied.[4]

The revised current theory is that patents are tools for economic advancement that should contribute to the enrichment of society through (i) the widest possible availability of new and useful goods, services and technical information that derive from inventive activity, and (ii) the highest possible level of economic activity based on the production, circulation and further development of such goods, services and information. Once patent rights have been acquired, the owners seek to exploit them in the marketplace. The possibility of attaining commercial benefits, it is believed, encourages innovation. But, after a certain period of time, these legal rights are extinguished and the now unprotected inventions are freely available for others to use and improve upon.

One common way to interpret the modern patent system is as a regulatory response to the failure of the free market to achieve optimal resource allocation for invention. As Geroski puts it:

> patents are designed to create a market for knowledge by assigning propriety rights to innovators which enable them to overcome the problem of non-excludability while, at the same time, encouraging the maximum diffusion of knowledge by making it public.[5]

This explanation for patents assumes that knowledge is a public good. This notion was nicely articulated by Thomas Jefferson who wrote that the 'peculiar character' of an idea is that 'the moment it is divulged, it forces itself into the possession of everyone, and the receiver cannot dispossess himself of it', and

---

into the Nature and Causes of the Wealth of Nations reprinted 1904, Edwin Cannan, ed., London: Methuen & Co., Book V, Chap. 1, Part III, 388; also see Bentham, J. (1962), 'A Manual of Political Economy', in *The Works of Jeremy Bentham* **31**, 71.

4 Mill, J.S. (1985), *Principles of Political Economy*, Book V, Chapter X, § 5, London: Penguin, pp. 295–6.

5 Geroski, P. (1995), 'Markets for technology: knowledge, innovation and appropriability', in P. Stoneman (ed.) *Handbook of the Economics of Innovation and Technological Change*, Oxford and Malden: Blackwell, pp. 90–131, at 97; also Arrow, Kenneth J. (1962), 'Economic welfare and the allocation of resources in invention', in *The Rate and Direction of Innovative Activity*, Princeton, NJ: NBER and Princeton University Press.

also that 'no one possesses the less, because every other possesses the whole of it'. He then went on to explain that 'he who receives an idea from me, receives instruction himself without lessening mine; as he who lights his taper at mine receives light without darkening me'.[6]

Such rights can be converted into market monopolies if the invention so protected results in a commercial product and depending on certain factors such as the relationship between the invention and the product, which may actually be protected by more than one patent. The public goods explanation for patents posits that the possibility of acquiring such rights encourages both investment in invention and the research and development needed to turn inventions into marketable innovations. Information about the invention as revealed in the patent and by the invention itself is diffused throughout the economy. In this context, it is helpful to conceive of a patent as a contract between the holder and the government on behalf of the citizenry. The holder receives an exclusive right over his or her invention in exchange for the payment of fees and – which is much more important – for disclosing the invention for others to learn from. Without a patent, the inventor would have no incentive to disclose it. This would be a loss for society if such lack of protection left the inventor with no alternative but to keep it secret. Such an alternative is a feasible option in several technological fields including biotechnology. But it is also true that many kinds of product would upon examination readily betray the invention that brought it into existence.

As for the creation of markets for knowledge, it might be useful here to explain why these are considered beneficial and how patents are thought to bring them into being. The explanation relates to the common situation that many patent holders are poorly placed to exploit their invention in the marketplace. Take the case of a creative but small company lacking the funds to develop and commercialise new products based upon its inventions. If such products are desirable for consumers, failure to commercialise would be a loss for society. But if the company owns a patent, a wealthier company may wish to license or buy the patent secure in the knowledge that the invention is legally protected. And if the invention were kept secret, how would bigger companies know about it? The disclosure of patent information makes it possible for prospective users to find inventions of interest and then to approach their owners.

The 'market failure' argument should not be carried to logical absurdities, and countries should not accept that it is necessary to allocate property rights on *every* intellectual output of the creator. Not everything created under the sun must be awarded intellectual property protection. Limits have to be placed on

---

[6] Letter from Thomas Jefferson to Isaac McPherson written on 13 August 1813, and reprinted in Andrew A. Lipscomb and Albert Ellery Bergh (eds) (1903), *The Writings of Thomas Jefferson*, vol. 6, pp. 330, 333–34.

the exact breadth of patent protection of inventions, and countries should take note of the effects of widening the current patent regime. In some instances, intellectual property rights can reach untenable levels whereby the intellectual property owner becomes a monopolist of discoveries or ideas, as opposed to 'inventions'. For example, if we were to allow every idea or discovery to be claimed and to be exhaustively protected under intellectual property rights, what of the future generations of inventors who would need to use such basic building blocks for further innovative activities? In such an instance, they would either have to incur licensing or other transaction costs to obtain permission to use these building blocks, or else they may attempt to work around the problem by attempting to disguise any appropriation of such blocks, incurring a potential cost of litigation. The final recourse would be to work around the protected building blocks and incur costly research expense. The costs of creating a new invention would increase, with detrimental effects on the rate at which new inventions are generated and reach the market. If patent protection were structured in such a manner as to unduly require the sacrifice of scarce resources, any societal benefits and economic rent accruing from a patent regime would be dissipated. These arguments are also the basis upon which a high level of inventiveness is required under patent law to ensure that commonplace, obvious or mere workshop inventions are not protected.

Thus, excessive protection will act as a disincentive to future creators, and certain types of basic building blocks of creativity must be left in the public domain. Where the line between the private and the public domains should be drawn is very difficult to determine but its ideal location will vary widely from one country to another and, one may argue, from one business sector to another. In countries where little inventive activity takes place, free access to technical information may well do more to foster technological capacity building than providing strong private rights over such information. In fact, technological capacity building may at certain stages of national development be best achieved by requiring foreign technology holders to transfer their technologies on generous terms rather than by trying to encourage domestic innovation by making strong legal rights available to all.[7] This suggests that developing countries should be careful not to make the rights too strong until their economies are more advanced. As we have seen, and have argued elsewhere, historical evidence indicates that several present-day developed countries, rightly or wrongly, took such a policy decision in the past.[8]

---

[7] This point applies to those developing countries that have attained a reasonable capacity to adopt and benefit from such technologies. Countries with very limited capacity have far less to gain from free access to technologies.

[8] Dutfield, G. *Intellectual Property Rights and the Life Science Industries: A Twentieth Century History*, Aldershot: Ashgate; Dutfield, G. and U. Suthersanen (2005),

2.2 TO PROTECT OR NOT TO PROTECT? MINOR INNOVATIONS AND PATENT LAW

The traditional justification of the patent system is that market power is conferred on an inventor for a limited duration for two purposes. First, the inventor is granted a reward for the introduction of technological developments by means of a buffer of protection against other competitors, and this guaranteed term of patent protection will act as a stimulus for further innovation. Second, societal welfare is assured as the inventor, in exchange for patent protection, will divulge his invention in the form of detailed patent specifications, thus ensuring a public knowledge base upon which future technological progress will be built.

The rationale for protecting minor innovations is tied closely to the patent system and its presumed inability to extend legal rights to innovations or discoveries that fall short of the inventive step and/or novelty bars. That is to say, limits are placed on the extent to which patent law will embrace inventive activity: only inventions which fulfil certain criteria will be protected, and adjunct to this axiom is the examination procedure which seeks to ensure that the patent system is not abused by the assertion of patents on spurious inventions. The following questions then arise, should we leave minor innovations or discoveries unprotected? Or should we lower the bars to incorporate them into patent law which is increasingly narrowing the scope of limitations and exceptions to the rights conferred and of safeguards against abuse of rights? Or else, should we seek an alternative means of protection? The next chapter considers options outside the patent system, primarily UMs.

'Harmonisation or differentiation in intellectual property protection? The lessons of history', *Prometheus*, **23**(2), 131–47.

3. Utility models and other alternatives to patents

Uma Suthersanen and Graham Dutfield

3.1 WHAT ARE UTILITY MODELS?

The term 'utility model' (UM) is a generic term which refers to subject matter that hinges precariously between that protectable under patent law and *sui generis* design law.[1] The term is also somewhat equivocal. It is not an accepted or clearly defined legal concept within the intellectual property paradigm. There is no global consensus on the term's meaning, due – it has been suggested – to there being 'fundamentally different concepts from one country to another'.[2] However, the term is bandied about by legislators and jurists to refer to a second tier patent system offering a cheap, no-examination protection[3] regime for technical inventions which would not usually fulfil the strict patentability criteria. This is an important factor: UM protection is accorded, cheaply and quickly, to inventions or innovations, many of which cannot gain protection under the patent regime.

If one examines national laws, one finds a plethora of terms used to mean 'utility model'. Thus, UM protection is referred to in Australia as 'innovation patent', in Malaysia as 'utility innovation', in France as 'utility certificate', and in Belgium as 'short term patent'. Some systems define utility models as incorporeal subject matter such as technical concepts or inventions or devices; while others anchor their definitions to three-dimensional forms. Yet others offer a utility model type of protection which, in actuality, is tantamount to patent protection without examination and for a shorter duration.

---

1 Suthersanen, U. (2001), 'Incremental inventions in Europe: a legal and economic appraisal of second tier patents', *Journal of Business Law* (July), 319–43.

2 AIPPI Resolution to Question 117, Montreal, June 1995.

3 There are always, of course, exceptions to the rule as, for example, evidenced by the Japanese utility model system which had an examination system from 1905. Tellingly, however, that characteristic of the regime was removed in 1992 by the Japanese Utility Model Law of 1959, as amended in 1995. See Macdonald, M. and Suthersanen, U. (1997), *Design and Copyright Protection of Products: World Law & Practice*, London: Sweet & Maxwell, p. JAPAN-9.

The confusion reflects the fact that within the international arena, a consideration of the nature and extent of protection under the various national 'utility model' laws reveals little consensus. Indeed, the UM law is not a standard feature within the intellectual property regime of many states. There are currently approximately 70 countries which provide, in some form or another, UM protection.[4] Included amongst the countries which do not have a UM regime are significantly the United States, United Kingdom and Canada. However, major industrial nations which have adopted the UM regime include Japan, South Korea and Germany.

As far as one can perceive, there are three traits common to all the national 'utility model' laws from a global perspective, which are:

1. All UM laws confer exclusive rights on the proprietor of the right (as opposed to an anti-copying right).
2. Novelty is a criterion in all UM systems, though the standard of novelty varies widely. It is noteworthy that national UM systems tend to adopt the International Patent Classification (IPC) as provided by the 1971 Strasbourg Agreement for the International Patent Classification, which facilitates the retrieval of patent documents in order to conduct effective novelty searches and determine the state of the art.
3. Registration is a requirement but usually there is no substantive examination of applications.

Even if one decides on the importance of the UM system, there is a further need to analyse the type of system required. This is very important. Among the EU member states, for example, one finds a wide disparity in the amount of applications made within each country, and this in turn correlates to the nature of UM protection. The major points of divergence can be summarised thus:

i. *Subject matter under protection.* Some UM laws protect only the three-dimensional form while others extend the umbrella of protection to cover technical inventions and processes. A majority of UM laws simply adopt the domestic patent law definition of protectable subject matter.

ii. *Granting procedure.* Many systems adopt a simple registration procedure with cursory examination; while a few implement a detailed examination process. In practice, some examining offices offer an optional detailed search facility with the payment of supplementary fees. Other jurisdictions

---

[4] See Table in Digest of National Laws in Macdonald, M. and Suthersanen, U. (1997), *Design and Copyright Protection of Products: World Law & Practice*, London: Sweet & Maxwell.

expressly call for a detailed search on validity to be carried out on the commencement of civil proceedings.

iii. *Substantive criteria.* Herein lies the greatest disparity between the UM systems. While all major UM systems adopt the criterion of novelty, the level of novelty required ranges from universal novelty, to relative novelty, to domestic novelty. A second criterion is usually, though not always, imposed in the form of inventiveness or of usefulness. Again, the standard employed for the level of inventiveness varies greatly. There is also a significant propensity within current UM laws to link the definition of the UM to an element of industrial application. A final element of divergence is the duration of protection which varies from six years to 15 years.

3.2 UTILITY MODELS IN THE WORLD

3.2.1 International Conventions and Multilateral Agreements

On the international front, UMs are recognised under the 1883 Paris Convention for the Protection of Industrial Property[5] as industrial property. However, the convention is silent on its definition and scope, and merely confirms that the international principles of national treatment and the right of priority is accorded to UMs.[6] Thus, Article 1(2) states:

> The protection of industrial property has as its object patents, utility models, industrial designs, trade marks, service marks, trade names, indications of source or appellations of origin, and the repression of unfair competition.[7]

The ambiguity of the term 'utility model' is also reflected in the cross-referencing and inter-dependency of priority periods between UMs, industrial designs and patents. Thus, a period of priority can be secured for an application for industrial design based on the filing date of a UM; and a period of priority can be secured for a UM application by virtue of a right of priority based on a patent application (and vice versa).[8]

---

[5] Revised at Brussels on 14 December, 1900, at Washington on 2 June, 1911, at The Hague on 6 November, 1925, at London on 2 June, 1934, at Lisbon on 31 October, 1958, and at Stockholm on 14 July, 1967.

[6] See Articles 2(1) and 4(C)(1), Paris Convention for the Protection of Industrial Property of 20 March, 1883 (Stockholm revision). The period of priority for applications for utility model is 12 months.

[7] Article 1(2), Paris Convention.

[8] Articles 4(E)(1) and 4(E)(2), Paris Convention.

The 1994 Agreement on Trade-related Aspects of Intellectual Property Rights (the 'TRIPS Agreement')[9] establishes minimum substantive standards for each of the major intellectual property regimes but fails explicitly to mention second tier or UM protection, thus leaving World Trade Organization (WTO) member countries free to formulate or reject second tier protection regimes as they see fit. Whilst there is no specific reference to UM protection under the TRIPS Agreement, it is arguable that by reference to Article 2(1), TRIPS Agreement, the relevant provisions of the Paris Convention provisions (including Article 1(2)) are extended to all WTO countries. But this still does not require WTO members or signatories to the convention to provide UM laws.

As we saw, national UM systems tend to adopt the International Patent Classification (IPC) as provided by the 1971 Strasbourg Agreement for the International Patent Classification, which facilitates the retrieval of patent documents in order to conduct effective novelty searches and determine the state of the art. Indeed, Article 1 states that the IPC covers not just 'patents for invention', but also 'inventors' certificates, utility models and utility certificates'.

Another significant agreement is the Patent Cooperation Treaty (PCT), the purpose of which is to facilitate patent applications in more than one country. By simplifying and cheapening the process, the treaty encourages patentees to secure protection over a broader geographical range. Instead of filing separately in all countries where protection is desired, applicants may file a single application in one language with a national patent office. When doing so they can designate all those signatory countries in which protection is also sought. After the examination of the patent, the application is transferred to one of nine International Search Authorities where a prior art search is conducted. After this it is then up to the patent offices of – or acting for – the designated countries to award the patent.

The facilitated means of securing international protection that the PCT provides for patents covers UMs as well. By virtue of Article 2, the PCT clarifies that 'application' means an application for the protection of an invention; references to an 'application' shall be construed as references to applications for patents for inventions, inventors' certificates, utility certificates, utility models, patents or certificates of addition, inventors' certificates of addition, and utility certificates of addition. In short, then, international applications may be for second tier patents as well as standard ones.

[9] Agreement on Trade-related Aspects of Intellectual Property Rights, Annex 1C of the WTO Agreement, 15 April, 1994.

3.2.2 Europe

Economic and innovation climate

Is there a link between innovation and utility models? The EU is characterised by substantial regional diversity in wealth, and competitiveness conditions differ substantially across regions. There are substantial national differences in innovation performance. Variations between member states are particularly high in relation to business R&D, high-tech patenting and the share of SMEs involved in cooperative innovation. The differences are greater in areas directly influenced by private decision making. In 2004, the European Commission published its summary of enterprise policy indicators in each of the 25 member states, summarising the areas of strength and weakness in each member state.[10] Countries which indicated innovation as one of their areas of strength included Finland, Denmark, Sweden and Germany. However, when we cross-refer this data to availability of second tier patent systems, we find that Sweden has no second tier system, whilst UM systems in Finland and Denmark are relatively new, having been introduced in 1991 and 1992 respectively.[11] On the other hand, the survey identifies countries with relatively longer experience of UM regimes such as Italy and Spain as having weak innovation systems.[12]

The European Innovation Scoreboard shows that the most innovative sector in the EU is electrical and optical equipment, followed by chemical and chemical products; the least innovative is textiles and textile products. Innovation in specific sectors tends to be member state specific – thus, the electrical equipment sector is most innovative in Finland while Germany leads in transport equipment. Interestingly, one of the factors responsible for the European utility model proposal was the European Commission's claim of a need for the protection of minor innovations by utility model in industries such as clock and watchmaking, optics, microtechnology and micromechanics. If we look at the European Innovation Scoreboard 2004, it indicates that there is an innovation gap between the United States and Japan, on the one hand, and the EU (as a whole) on the other.[13] Although the EU innovation performance has been constant since 1996

---

[10] A Pocketbook of Enterprise Policy (2004), 'How member states and candidate countries rank in the 2004 enterprise policy scoreboard', accessed at http://europa.eu.int/comm/enterprise/enterprise_policy/competitiveness/doc/pocketbook_2004.pdf.

[11] Finnish Law No. 800 of 10 May, 1991; Danish Act No. 130 of 26 February, 1992.

[12] Italian Law on Industrial Model Patents, Royal Decree No. 1411 of 25 August, 1940; Industrial Property Statute of 1929.

[13] European Innovation Scoreboard (2004), 'Comparative analysis of innovation performance', commission staff working paper [Sec(2004) 1475], accessed at http://register.consilium.eu.int/pdf/en/04/st15/st15189.en04.pdf.

and has not faltered, innovation performance in the United States and Japan has improved, thus widening the gap. An interesting element of the research is that the gap is explained away by comparing three factors between the United States and the EU:

- patents (50 percent of the gap)
- working population with tertiary education (26 percent)
- R&D expenditures (11 percent) – mainly business R&D.

Public sector spending on R&D is comparable in the EU, the United States and Japan. The various reports from the European Commission lament the fact that there is, despite the increase in business R&D in several EU member states, a gap with the United States and Japan in respect of patenting. The major worry is that United States high-tech patenting in Europe is about seven times higher than European patenting in the United States; the situation with Japan is almost as unbalanced, with a strong position of Japanese high-tech patenting in the United States.[14] This is in contrast to the situation in individual member states. Denmark, Finland, Sweden, Germany, France and the Netherlands record a higher percent of GDP in public sector R&D expenditure than the United States. These member states also perform well in patent registrations, while in the case of Ireland the share of high-tech products in total exports is in excess of 40 percent, compared to around 30 percent in the case of the United States.

If we accept that intellectual property registrations are an indicator of innovation activities, do we also correlate this with the fact that the United States has no UM system? As the studies discussed in Chapter 3.6 show, the patenting surge in the United States was driven by several factors:

- changes in the management of R&D activities
- smaller and medium sized businesses and less established patentees were aggressively exploiting the patent system
- the amount of United States domestic patent applications granted by USPTO declined very little (with suggestions of regulatory capture of the patent office)
- the low patentability criteria.

The studies also are doubtful whether the actual intensity of research effort in the United States has risen during the same period.

---

[14] Annual reports by the Commission, 1991–2003.

Utility model laws in individual member states

Standard patent protection within the EU is available either under national patent systems or under the European Patent Convention, with member states' patent laws being, to a certain extent, harmonised by the accession of all EU states to the European Patent Convention. The European Patent Convention does not create a uniform right but it does provide the applicant with protection in as many of the signatory states as he wishes, and rationalises the grant of patents in Europe by means of a centralised application and examination procedure managed by the European Patent Office in Munich. In this manner, the European Patent Convention fulfils the function of a harmonising patent regime within the EU.

There is currently no consistent policy for UM protection across EU member states. Table 3.1 offers a bird's eye view of the types of second tier protective regimes in the major EU member states. None of these systems conform to a single 'utility model' system. Missing countries include the United Kingdom and Sweden. While Sweden does not have any other means of protection of innovation other than under the patent and design laws, the United Kingdom has a second means of protection via its unregistered design right system.

Table 3.1 Current second tier protection regimes in major EU member states

Belgium	Brevet de courte durée/Octrooi van korte duur
Denmark	Brugsmodel
Germany	Gebrauchsmuster
Greece	Πιστοποιητικό νποδειγματοζ χρησιμοτηταζ
Spain	Modelo de utilidad
France	Certificat d'utilité
Ireland	Short term patent
Italy	Brevetto per modelli di utilità
Netherlands	Zesjarig octrooi
Austria	Gebrauchsmuster
Portugal	Modelo de utilidade
Finland	Nyttighetsmodellagen

If we reduce the laws into key elements and categorise them according to their common denominators, for example subject matter protected, conditions of protection, examination and length of protection, we arrive at a different classification as set out in Table 3.2 – that there are three separate systems at work here. The latter table reflects, for instance, the fact that French and Belgian law, despite terminology, are not really UM systems but quick 'reservation' systems for patent applications, and offer protection akin to the standard patent law. The

Table 3.2 Classification according to characteristics of the law

Three-dimensional regime	German regime	Patent regime
Italy Denmark Finland Greece Portugal Spain	Germany Austria EC Proposed Utility Model Directive	Belgium Ireland Netherlands France

Italian system, however, despite the usage of the word 'patent', is more akin to the traditional UM regime.

The patent classification

Table 3.3 shows that the so-called UM laws in four member states mimic the domestic patent laws, with identical requirements for protection. The criteria of novelty and inventive step are identical to that under patent law. Thus, the full inventive step criterion and absolute or international novelty are required. There is no three-dimensional or form limitation within the definition.

Table 3.3 Patent type regimes in Netherlands, Belgium, Ireland and France

Countries	Netherlands, Belgium, Ireland, France
Subject matter	Similar to patent law i.e. any invention except those specially stipulated (identical to Articles 51–53, European Patent Convention)
Criteria of protection	Novelty, inventive step, industrial application. Standards of novelty and inventive step are identical to those under patent law
Duration	6–10 years
Examination	No examination except as to formalities

As stated before, an example of this regime is the French *certificat d'utilité* which runs parallel to the patent law system with identical provisions and conditions of protection applying to both regimes.[15] The only main difference is that

15 Article L 611–2 of the Intellectual Property Code, Law No. 92–597 of 1 July, 1992.

inventions are granted a short term protection of six years under a *certificat d'utilité,* and the grant of *certificat d'utilité* is made without a prior search report.[16] Another variation on this patent prototype theme is the Dutch *zesjarig octrooi* which is governed by the Netherlands Patents Act and provides for a fast-track, no-examination registration system whereby the applicant may obtain a short term patent with a duration of six years from the application date after formal examination. It is open to the applicant to opt for a full patent registration or to remain with this short term registration. The choice must be indicated at the application stage with the applicant requesting a novelty search report if he wishes a full 20-year patent.

The three-dimensional classification

Table 3.4 demonstrates countries which offer the traditional type of UM system i.e. the protectable invention must be embodied in a three-dimensional form. The conditions of protection are generally also much less stringent than that under patent law. Thus, the standard of inventiveness is much less, and can be termed a diminished inventive step requirement. The standard of novelty differs within all these systems.

In Spain, UMs are granted protection under the Spanish Patents Act and are defined as inventions which consist of a form, structure or constitution which results in a practical and appreciable advantage for its use or manufacture and in particular utensils, instruments, tools, apparatus, devices or parts thereof.[17] UMs are truly considered as minor inventions, with lower thresholds being required than that for patent protection. Novelty is judged at a local level with the state of prior art consisting of everything which has been published in Spain by any means. There must also be some utility or advantage being offered. Other EU member states which adopt the three-dimensional paradigm include:

- Denmark – offering protection to any creation which may be industrially applied and which contains the solution to a technical problem.[18]
- Finland – according protection to any new invention embodied in a shape or design of a device or their combination which is commercially exploitable. Protection is for ten years following a formalities-only examination. An option for a substantive examination is available on payment of a fee.[19]

---

16 Despite the high protection thresholds, there is only an examination as to formalities. A similar nature of protection is also available under the Belgian Patent Law of 28 March, 1984.

17 Spanish Patent Act of 20 March, 1986.

18 Danish Act No. 130 of 26 February, 1992.

19 Finnish Utility Model Act 800/91 of 1 January, 1992.

- Greece – granting protection to any invention which is novel, industrially applicable and possesses a solution to a technical problem.[20]
- Italy.[21]
- Portugal – according protection to all models which are new, industrially applicable, functional and consisting of inventive step. Although there is a detailed examination, this has no effect in determining questions of validity in the event of future civil proceedings.[22]

Table 3.4 Utility model-type regimes in Italy, Spain, Denmark, Finland, Greece and Portugal

Countries	Denmark, Finland, Greece, Italy, Portugal, Spain
Subject matter	Three-dimensional inventions defined in terms of 'form', 'structure' or 'configuration' which results in a 'practical and appreciable advantage' for their use or manufacture and in particular utensils, instruments, tools, apparatus, devices
Criteria of protection	The criterion of novelty varies from local novelty (in Spain and Italy) to regional or absolute novelty (in Finland, Denmark and Greece). Sometimes, the statutory requirements are at variance with actual standards applied. There is no consensus as to whether there should be a second criterion, and if so, whether it should be 'inventive step' or 'utility' or 'advantage' in a particular field. Industrial application is a requirement
Duration	7 years – Greece; 10 years – Denmark, Spain, Finland, Italy; 15 years – Portugal
Examination	No examination except as to formalities

3.2.3 Germany

Economic and innovation climate in Germany

Germany had no coherent or unified industrial property policy until the creation of the German free trade area in the late nineteenth century. The first patent law

20 Greek Utility Model Law 1733/87, 1 January, 1988.

21 Italian Law on Industrial Model Patents Royal Decree No. 1411 of 25 August, 1940.

22 Portuguese Industrial Property Code (Law No. 16/95), 1 June, 1995.

in the German Reich was introduced relatively late in 1877 against a background of anti-monopoly and anti-patent movements. Part of the reason for this push for industrial property protection was the change in the economic landscape of Germany. German territories, especially Prussian ones, were agrarian at the beginning of the nineteenth century. Between 1850 and 1870, however, Germany was industrialising rapidly and industrial leaders such as the Siemens firm organised pro-patent lobbies. Today, in terms of the number of patents, Germany belongs to the leading international group. German patent output increased in recent years after an initial post-reunification downturn. In addition to higher innovation output, this general growth in patent activity may be at least partly due to companies' increasing strategic focus on legal protection for intellectual property on globalised markets.

A Deutsche Bank Report identified the following facets within Germany's economic and industrial structure:

- Germany has high labour costs coupled with too few other highly skilled personnel.
- There is a lack of innovative drive, and a reluctance to invest by SMEs – indeed, the R&D expenditures by Sweden and Finland now far exceed the German level.
- The established companies are now more likely to be the major contributors towards innovation, rather than SMEs, and they generate a far larger proportion of their turnover with innovative products.
- Leading German industries are the automobile industry, mechanical engineering, medical, precision and optical instruments, measurement and control engineering, textiles (technical textiles) and parts of the chemical industry.

The report also identified that, on average across industries, German innovators first take advantage of their time lead and secrecy, with patenting coming only third. Another tactic employed by the industry is to incorporate deliberately complex technical product design, so as to discourage reverse engineering, and to use trade marks.[23] The important factor is that the German local market and industry is considerably knowledgeable as to the intellectual property regime.

Current substantive law

The German UM ('Gebrauchsmuster') system has been in place since 1891. It was introduced because it was perceived that patent law was unsuitable for some types of inventions. This derived from the stringent German patentability re-

---

23 Deutsche Bank research report on innovation, 10 October, 2003.

quirement that inventions represent a *technischer Fortschritt*, or 'technical step forward in the art', a standard that was considered too elevated to be met by minor inventions.[24] Thus, the UM system was introduced with a lower standard of inventiveness, a non-examination system and a short period of protection. Conceptually, the German system was not a supplemental patent regime; rather the UM protection was introduced as a supplement to 1876 legislation protecting copyrights and designs. The German UM regime was a classic UM regime i.e. one that is tied inextricably to design protection and to preventing third parties from duplicating the external configurations of certain handtools and other everyday implements whose creative contribution fell chronically short of the inventive height that the mature patent paradigm required.[25] Thus, the classic UM regime, as exemplified by German legislation, was originally conceived as a form of design protection. [26]

The UM system was, and is still, considered to be an indicator and encouragement of local innovative activity in the fields of agricultural products and domestic appliances. Thus, historically, there has always been a 'three-dimensional' requirement for UM laws, as such a requirement related to the types of innovative activity to be encouraged. One could almost say that this was an early registered design system for functional designs. What the following analysis reveals is that the German local industry is very knowledgeable about the system and utilises it to its fullest extent.

The German Utility Model Law, last amended on 21 January 2005, protects any inventions of technical character that are new, based on inventive step and are capable of industrial application.[27] In addition to the usual excluded subject matter under patent law (such as discoveries, scientific theories, aesthetic creations, etc.)[28], the UM law additionally excludes inventions relating to processes and biotechnological inventions. Furthermore, as opposed to German patent law, UM law requires lesser thresholds of protection. The subject matter of a utility model is considered new if it does not form part of the state of the art. The state of the art comprises any knowledge made available to the public by means of a written description or by use within the territory of Germany before

---

24 Janis, M.D. (1999), 'Second tier patent protection', *Harvard International Law Journal*, 40, p. 151.

25 Ibid., p. 159.

26 Krasser, p. 953; Stephen P. Ladas, *Volume 2 Patents, Trademarks, and Related Rights: National and International Protection* § 551 (1975), p. 952; Janis, Mark D. (1999), p. 159.

27 Article 1(1), German Utility Model Act of 28 August, 1986, as amended on 21 January, 2005 (*Federal Law Gazette I*, p. 146).

28 The provisions on excluded subject matter under the German utility model and patent laws are based on Articles 52(2), (4) and 53, European Patent Convention.

the date relevant for the priority of the application (i.e. 'local novelty'). Non-obviousness under utility model law is easier to meet than under patent law as the utility model law refers to 'inventive *step*', while the patent act requires 'inventive *activity*', which is a higher level of inventiveness.[29] The process of obtaining a utility model registration is made even easier by the fact that there is no pre-grant examination.

The Austrian law is virtually the same as the German law, though it goes further in allowing utility model protection to extend to the underlying algorithm or process of a computer program; the only exclusions under Austrian law seem to be microorganisms and similar biotechnological inventions, and even chemical processes are protectable by way of a utility model.[30]

Empirical evidence

The German utility models system continues to be a popular one. Because the utility model registration is issued speedily, inventors tend to file for patents and utility models simultaneously. The result is that nearly one German patent application in every two is accompanied by a utility model registration.[31]

Figures on utility model applications filed at the German Patent and Trade Mark Office, which include PCT filings, show that applications increased steadily up to 1999. The numbers have fallen from their 1999 peak, but not by a great deal, and the drop is insufficient to indicate that use of the system is declining. Admittedly, the total of utility models in force has fallen from 115,535 in 2000 to 108,175 in 2003, but given that the number of 2002 and 2003 applications are both greater than in 2001, this may be a short term trend.

The fact that the German utility model system continues to be popular with industry despite Germany's being ranked by the European Commission as sixth most innovative economy in the world,[32] suggests that the advantages of utility models are not confined to that of facilitating an economy's advancement from developing to developed country status. But it may not be quite as simple as that given that the poorer East Germany was incorporated into the Federal Republic of Germany in 1990. Moreover, if we compare Germany with Japan we

---

[29] German Utility Model Act, Art. 1(2) and 3(1). The terminology can be confusing. German patent law requires *erfiinderische Tatigkeit* whereas a utility model requires only *erfinderischer Schritt* – the normal English translation of the former is inventive activity and of the latter is inventive step. This is confusing as French and English wordings of patent law tend to use the phrase 'inventive step' as the equivalent to the first of these German terms.

[30] Article 1(2), Austrian Utility Model Law, 1 April, 1994.

[31] Kaufer, E. (1989), *The Economics of the Patent System*, Routledge, p. 12.

[32] European Innovation Scoreboard 2004. Comparative Analysis of Innovation Performance. Commission staff working paper [Sec(2004) 1475].

find the two countries, despite their similar economic circumstances and innovation climate, diverged from the mid 1990s, with the German filing numbers remaining high while Japanese filings markedly declined.

However, it does appear that despite the popularity of utility models in Germany, SMEs are still reluctant to invest into R&D. It is argued that Germany's leading position in the world market for products is a result of its *Mittdstand*, a network of SMEs, including family firms; investment levels according to the report have been steadily dropping since 1998; so worrying is this trend that the German government has started a seed fund to assist R&D-based start-ups.[33] Arguably, the utility model system in Germany may not really be serving one of its primary purposes which is to spur innovation through SMEs.

3.2.4 Asia and the Pacific

A more detailed study is made of the Asia-Pacific region in Part Two. Nevertheless, it is pertinent to have a quick overview of the Asian region. Tables 3.5 and 3.6 reveal that many countries (though notably excluding Singapore) have adopted a utility model law. However, there is great difficulty in classifying these laws since they vary greatly.

3.2.5 Africa

Annex II of the 1999 Revised Bangui Agreement of 2 March 1977, on the Creation of an African Intellectual Property Organization (OAPI) is much more detailed than, say, the utility model provisions of the Andean Community Decision 486. It also establishes unitary intellectual property systems so that rights granted have automatic effect in the territories of every member state, albeit subject to the relevant countries' national legislation.

Accordingly, protected utility models 'shall be implements of work or objects to be utilized or parts of such implements or objects in so far as they are useful for the work or employment for which they are intended on account of a new configuration, a new arrangement or a new component device, and are industrially applicable'. Novelty is territorial so that an application would fail only if the implement or object had already been publicly used or disclosed in print in the territory of a member country. The term of protection is ten years from filing date. As in the Andean Community countries, utility model applications can be

---

33 The OECD Observer (September 2005), 'German Innovation', article summarising the results of an in-depth report on innovation in Germany in Röhling, K. and T. Multhaup (2005), 'Innovative SMEs in Germany', in *OECD SME and Entrepreneurship Outlook*, Paris: OECD.

Table 3.5 Utility models in Central and Far Eastern Asia and Pacific region (ex ASEAN countries)

Australia	Patents Act 1990, amended by Patents Amendment (Innovation Patents) Act 2000. Second tier protection comes within the innovation patent regime, which has no substantive examination prior to grant. The requirements are universal novelty and an innovative step – which is a lower standard of inventive step. Term of protection = eight years
China	Patent Law and its Implementing Regulations, 1 April 1985, revised 1 January 1993. Under Chinese law, utility models are defined as 'any new technical solution relating to the shape, structure, or their combination, of a product, which is fit for practical use'. There is no substantive examination. Term of protection = ten years
Hong Kong	Patents Ordinance 1997, Cap. 514, Part XV, s. 108 *et seq.* Short term patents protect a product or an industrial process which is new, inventive and has industrial application. No substantive examination. Term of protection = eight years
Japan	Law No. 123, 13 April 1959, as last amended on 12 June 1996, w.e.f. 1 April 1997. Utility model protection granted to devices which are new and industrially applicable and which relate to the shape or construction of articles. Term of protection = six years
Kazakhstan	Criminal Code, 1 January 1998, which is the basis of intellectual property protection. The Criminal Code establishes that the following are criminal acts: the illegal use of a utility model; the unauthorised disclosure of the essence of a utility model before such information is officially made public; the misappropriation of authorship of an invention
Korea	Utility Model Act, 31 December 1961, as amended 1 July 1999. The utility model must be novel, capable of industrial applicability and possess an inventive step. Examination is only with respect to formalities. Term of protection = seven years
Kyrgyz Republic	Law on Temporary Provisions Concerning Industrial Property of the Kyrgyz Republic, 2 August 1993. 'Utility models' are defined as new and industrially applicable features used in the production of consumer products or components of such products. The conditions for registration are novelty and industrial applicability; the utility model must also possess a functional and inventive char-

Table 3.5 (continued)

	acter. Domestic novelty is required. The application is examined *ex officio* for formalities only. Term of protection = eight years
Macao	Macanese Industrial Property Act 1999. Utility models are models of objects or parts of objects (appliances, instruments, tools, containers etc.) made for practical use, which, by means of a new shape or configuration or a new mechanism, increase or improve the handling of such objects. Utility models must be three-dimensional. Novelty, inventiveness and practical applicability are also required in the case of utility models, though the level of inventiveness is low. Term of protection = 15 years
Taiwan	Patent Law (Chapter III: New Utility Model Patent), 29 May 1944; on 21 January 1994. A 'new utility model' is defined as a creation or an improvement which has been made in respect of the form, construction or fitting of an object. The utility model must be novel, capable of industrial applicability and involve an inventive step. Examination is substantial. Term of protection = 12 years
Turkey	Decree Law No. 551 on the Protection of Patent Rights, 27 June 1995, amended by Law No. 4128 of 7 November 1995. The law provides that inventions which are novel and applicable in industry may be protected by utility model certificates. The two conditions for registration are novelty and industrial applicability. Local novelty is required. Examination is for formalities only. A report as to the state of art, available under the patent system, will be drawn in respect of a utility model only on request of the applicant. Term of protection = ten years
Uzbekistan	Rules of Preparing and Filing Applications for Inventions and Utility Models, effective 22 September 1992. 'Utility models' are defined as new and industrially applicable features used in the production of consumer products or components of such products. In order to be registered the utility model must be new and capable of industrial applicability. Novelty is determined at a universal level. The application is examined *ex officio*. The examination is cursory. The result of the examination is taken into account in determining questions of validity in the event of opposition, invalidation or infringement proceedings. Term of protection = eight years

Table 3.6 Utility models in ASEAN countries

Country	Name	Novelty	Inventive step	Subject matter of protection	Substantive examination	Protection period
Cambodia	Utility model	Same as patents	Not required	Same as patents	No	7 years
Indonesia	Petty patent	Same as patents	Not required	Same as patents	Only upon request	10 years
Malaysia	Utility innovation	Same as patents	Lower standard than patents	Similar to patents	Yes	15 years
Philippines	Utility model	Local novelty only	Not required	Any technical solution of a problem in any field of human activity	No	7 years
Thailand	Petty patent	Same as patents	Not required	Same as patents	Only upon request	10 years
Vietnam	Utility solution	Same as patents	Not required	Same as patents	Only upon request	10 years

converted upon request to patent applications as well as vice versa. Searches are carried out for novelty and industrial applicability. There is also an examination of applications to ensure that the subject matter does not fall under the stated exclusions, that the applications are 'restricted to a single principle subject', and that 'the claim or claims defining the scope of the protection sought' do 'not go beyond the contents of the description'.

The African Regional Industrial Property Organization (ARIPO) Protocol on Patents and Industrial Designs (Harare Protocol)[34] defines utility models as 'any form, configuration or disposition of elements of some appliance, working tools and implements as articles of everyday use, electrical and electronic circuitry or other object or part thereof in so far as they are capable of contributing some benefit or new effect or saving in time, energy, or labour or improving the hygienic or sociophysiological working conditions by means of new configuration, arrangement or device or a combination thereof and are industrially applicable'. The ARIPO office carries out a substantive examination of applications for novelty and industrial application. Utility models need only be new in the designated countries. Again, applications can be converted upon request to patent applications as well as vice versa.

3.3 UTILITY MODELS IN THEORY AND PRACTICE

Utility models can be justified on both theoretical and practical grounds. These are closely related. The theoretical rationale for UMs derives from the facts that most social welfare-enhancing inventions are cumulative in nature and that a great deal of them are subpatentable in the sense that the novelty and inventive step requirements are too high for the patent system to accommodate them. In fact, in today's industrial society different levels of innovative activity apply in different areas. Thus, major technological breakthroughs may be more common in some industrial sectors (e.g. biotechnology and ICT) than others (e.g. electronics), but are hardly everyday occurrences in any of them.

Although utility model laws confer exclusive property rights, the underlying rationale is usually to accede to industries' call for an anti-copying right or a misappropriation tort. Indeed, the main practical justification derives from the fact that many inventions are vulnerable to unfair copying, and that in many cases, the sub-patentable ones are the most vulnerable of all. If one accepts the 'unfair copying' argument favouring IP protection, it follows that any subject

---

34 Adopted on 10 December, 1982 at Harare (Zimbabwe), and amended by the Administrative Council of ARIPO on 11 December, 1987, 27 April, 1994, 28 November, 1997, 26 May, 1998, 26 November, 1999 and 30 November, 2001.

matter evincing some sort of intellectual or capital investment, and which is open to imitation and copying, should be considered a worthy intellectual property good and requiring protection. Indeed, some products, once invented and marketed, are especially susceptible to being copied at negligible or zero cost. Often, these products come from industries whose innovations tend to be small, incremental and cumulative in character. Evidently, it is for these industries that a UM system is likely to be most beneficial.

However, whilst some industries tend to be very enthusiastic about low cost, fast protection regimes (such as a no-examination utility model system or a no-registration property right), other industries are highly suspicious of such systems, especially when they are viewed as curtailing industries' right to innovate on the basis of low-level access to a large public domain and 'creative imitation', a term adopted to show that imitation is often not only an essential stage in learning to innovate but can even be creative in itself. So one really needs to know to what extent copying is a problem in the different industries, and whether such inventions are better left in the public domain and open to imitation and free riding.

Apart from these issues, consideration of the practical advantages of UM regimes requires an understanding of how firms compete in an innovative environment. One means for investigating this issue may be to utilise Howard and Guile's model for market competitiveness. In relation to successful commercialisation of any product, they point to three distinct driving forces:

i. technology-enabled commercialisation
ii. design-driven commercialisation
iii. market-driven commercialisation.[35]

The technology-enabled commercialisation is driven by a revolutionary scientific or technical discovery, with little emphasis or understanding of the market for resulting products of the technology. Such revolutionary technology replaces old ways with completely new products and processes. But as Howard and Guile warn us:

> Technology-enabled commercialisation … is a very risky undertaking … A major industrial nation cannot depend principally on technical revolutions. They occur infrequently, entail a high level of risk that makes them haphazard, and they do not provide the sustained push that an economy must have to endure in the market marathon.

[35] Howard, W.G. Jr. and B.R. Guile (1992), 'Profiting from innovation,' in N. Rosenberg, R. Landau and D. Mowery (eds), *Technology and the Wealth of Nations*, Stanford, CA: Stanford University Press, p. 397, *et seq.*

Instead, the more frequent types of commercialisation are design-driven and market-driven commercialisation:

> Once a concept has been proven in the market, competition shifts to price, performance, and design features, with less competitive emphasis being put on fundamental differences in products or service concepts.

With the proliferation of other competitors within the same product market, the emphasis shifts to intra-product competition. The ability of the product to succeed is dependent on the product design and production process, especially in relation to the product cycle time – i.e. the need to continually update a product in record time. The third stage is market driven, whereby it is the market's need within a particular product market which drives the requirement to innovate further.

The last two stages of design and market-driven commercialisation correlate with the reasons why countries introduce the second tier patent system:

- The majority of inventions today are incremental in nature which, though novel, exhibit a modicum of inventiveness.
- The majority of innovators in most economic infrastructures receive no encouragement or incentive for continued development.
- These incremental products are open to imitation when their product is placed on the market.

There is a view that utility models are especially good for relatively innovative developing countries seeking to advance technological capacity through local incremental innovation. For example, Juma put forward five reasons why utility models are appropriate for such nations.[36] The first is that they enable artisans to secure protection for innovations that do not meet the stricter novelty and inventive step requirements of patent law. Second, they make it possible to increase the role of small-scale innovators and artisans in economic development and help them stay in business in the face of new technologies that might threaten their livelihoods. Third, they act as a spur to enhanced levels of innovation. Fourth, they are cheaper to acquire than patents. And finally, they may become a source of data on innovative activity and experience in technological management.

As for empirical evidence, Kumar[37] found that in the East Asian countries he studied (i.e. Japan, South Korea and Taiwan), a combination of relatively weak

---

[36] Juma, C. (1989), *The Gene Hunters: Biotechnology and the Scramble for Seeds*, Princeton, NJ: Princeton University Press, pp. 231–2.

[37] Kumar, N. (2002), *Technology and Economic Development: Experiences of Asian Countries*, London: Commission on Intellectual Property Rights.

IPR protection and the availability of second tier patents like utility models and design patents encouraged technological learning. The weak IPRs helped by allowing for local absorption of foreign innovations. The second tier systems encouraged minor adaptations and inventions by local firms. Later on, the IP systems became stronger partly because local technological capacity was sufficiently advanced to generate a significant amount of innovation, and also as a result of international pressure. India's experience is somewhat similar, except that no second tier protection was provided. This did not hurt the chemical or pharmaceutical industries, but may, he suggests, have hindered the development of innovative engineering industries.

It is often claimed also that UM systems are particularly advantageous for small and medium sized enterprises (SMEs). For one thing, it is quite likely that SMEs have a large presence in those industries where minor innovation is the norm, and unfair copying is rife. Indeed, it is often argued that a cheap and rapid second tier patent regime would improve the legal environment for SMEs, especially those which are engaged in an ongoing process of innovation and adaptation. This is more so in relation to certain types of product sectors which are concerned, not so much with revolutionary technological breakthroughs, but more so with incremental or improvement innovation. For example, one reason for the Proposed European Commission Utility Model Directive is the perceived need for a rapid and cheap protective regime for such minor innovations in the following industries: toy manufacturing, clock and watchmaking, optics, microtechnology and micromechanics.

For another, it may even be that more innovations, both of the breakthrough and incremental varieties, emanate from SMEs than from larger multinational conglomerates. If this is so, it is important to gauge whether the current patent regime is attuned to the needs of SMEs and the types of inventions they produce. Many inventions which originate in SMEs have a lower standard of inventiveness, and are prime candidates for free-riding activities by competitors. Consequently, UMs may be highly pro-innovation and consequently good for the national economy.

Another reason why UMs may be good for SMEs is that the cost factor may inhibit them from using the patent system as much as they would desire. The second tier patent regime is viewed as the ideal solution as it is a system geared towards the needs of SMEs.

On the other hand, it may also be argued that the rationale for UM systems is inherently unsound. There are fundamental problems with UMs as a policy solution to the question of what, if anything, should be done about sub-patentable inventions. First, the fact that the UM regime represents a lowering of thresholds without an appropriate examination system in place may result in legal uncertainty and excessive litigation. Indeed, there is a reasonable concern that larger market players may use UMs as a means of circumventing the more stringent criteria under the patent system and overuse the system in ways that make it hard for SMEs to compete. Certainly, the lack of substantive examination prior to grant will give rise to uncertainty for third parties when conducting infringement searches to ascertain what valid rights exist in a particular field of technology, which may act as an additional barrier to competitors. It can also lead to abusive behaviour. In China, for example, it is often alleged that UMs are granted to local 'inventors' for inventions imported from overseas and that the local owners sometimes assert their rights against the original inventors.

3.4 COSTS AND BENEFITS OF UMs

What are the benefits to be gained from a UM system?

- UMs encourage production of more intellectual property goods, and in particular encourage local innovation.
- If the region is a net importer of intellectual property goods, extended protection may be beneficial to encourage local industry to produce more goods, and thus decrease imports.
- UMs provide protection of intellectual property goods which currently cannot be protected under the current intellectual property regime (for example, patent, design and copyright laws).
- UMs prevent free-riding from other predatory firms which expend no R&D costs or investment.
- They can provide revenue to governments in the form of registration, search, publication fees etc.
- They provide a source of information via published specifications.

This should be contrasted with the costs of the UM system:

- Possible increase in litigation
- Possibility of economic rent-seeking behaviour i.e. re-direction of funds in an effort to gain UM protection
- Possible increase in net royalty and licensing fees to overseas producers
- Potential cordoning-off of areas of research using patents and UM protection.

It may also be argued that the rationale for UM systems is inherently unsound. Objections relate to the fact that in essence, the choices facing policymakers in respect of sub-patentable inventions are far from straightforward. They may choose to leave them unprotected, to lower the inventive step threshold so that they become patentable, or they may seek to create alternative legal means of protection. Alternatives include introducing a liability rule-based system such as a statute-based tort or misappropriation law, or a hybrid property rights system such as industrial design rights or utility models. These systems could operate in place of patents or copyright, as the case may be, or could alternatively operate in parallel, so that patents and copyright could then be used only for those inventions and creative works displaying a demonstrably high inventive step or level of originality.

There are more fundamental problems with UMs as a policy solution to the question of what, if anything, should be done about sub-patentable inventions.

3.4.1 Uncertainty and Unfairness

In traditional patent law, a rigorous and lengthy examination procedure is imposed to allow the relevant patent office to determine whether the invention is patentable. In this manner, the examination process ensures that patents are awarded to inventions which demonstrate the requisite level of novelty and inventiveness. Partly in deference to the high societal costs imposed by the conferral of a proprietary right which will stifle competition for a short duration, and partly to forestall duplication of patent protection for identical or similar inventions,[38] the examination system is a major element of the patent regime.

The fact that the UM regime encourages a lowering of thresholds without an appropriate examination system in place may result in legal uncertainty and excessive litigation. Indeed, there is a reasonable concern that larger market players may use UMs as a means of circumventing the more stringent criteria under the patent system and overuse the system in ways that make it hard for SMEs to compete. Certainly, the lack of substantive examination prior to grant will give rise to uncertainty for third parties when conducting infringement searches to ascertain what valid rights exist in a particular field of technology, which may act as an additional barrier to competitors.

It is difficult to test the validity of such a concern, but it is noteworthy that a survey carried out on behalf of the European Commission in 1993 found that on average 50 percent of industry in the United Kingdom, France, Spain, Italy and Germany was 'partly to greatly dissatisfied by the application of the UM system in Europe at present' with the satisfaction being greater amongst SMEs. The causes of this dissatisfaction were not totally specified but the fact that dif-

[38] Examination procedures are a relatively new concept with examinations of patent applications for novelty having begun in the United Kingdom in 1905, as a result of the Fry Committee recommendations that evidenced that approximately 40 percent of patents were granted for inventions already published in earlier United Kingdom patent specifications.

ferent standards exist between various countries was clearly a contributing factor in addition to any discontent that might exist about the specific features of protection in any particular country. Thus, 76 percent of the SMEs surveyed expressed a high interest in the adoption of a community wide UM system whereas interest among larger companies was much smaller.

3.4.2 Is Copying Always Bad?

Hybrid systems like UMs tend to rely extensively on the argument that cumulative innovations are vulnerable to unfair copying which should not be allowed. This argument is anchored to the classical natural rights justification of the intellectual property regime, that protection is conferred in order to enable inventors and creators to reap their just rewards by preventing misappropriation; hence, such innovations should be protected. Moreover, it is traditionally thought that protection against misappropriation is important to SMEs to encourage small-scale innovations (this is discussed in section 3.5.2)

However, herein lies a danger. Intellectual property rights are predicated on the notion that inventions and creative works are in the public domain *except* such works that fulfil the traditional standards of novelty, inventive step, originality or distinctiveness. If we provide second and third tier rights for inventions and works that fail to meet these standards, thereby allowing insufficiently inventive or original works to get protection by the back door, we undermine both the public–private boundary and the integrity of existing IPRs and their doctrines.[39] After all, in a market-based economy it is generally accepted that all market actors, including competitors, follow-on creators and consumers, should be allowed to freely use any work which falls short of the required standards.

Indeed, as some courts and jurists have argued, copying and free-riding is necessary, if not beneficial, for competition.[40] As we saw, imitation is an essential stage in learning to innovate and can even be creative in itself.[41]

---

[39] By introducing parallel regimes of protection, we erode established doctrines in copyright and trade mark laws such as the 'functionality', 'separability' and the 'idea-expression' – Suthersanen, U. (2003), 'The European Court of Justice in Philips v Remington – trade marks and market freedom', *Intellectual Property Quarterly*, **3** (3), 257; Rushton, M. (2001), 'An economic approach to copyright in works of artistic craftsmanship', *Intellectual Property Quarterly*, **3**, 255.

[40] Dam, Kenneth W. (1995), 'Some economic considerations in the intellectual property protection of software', *Journal of Legal Studies*, **24**, 321.

[41] Dutfield, G. and U. Suthersanen (2005), 'Harmonisation or differentiation in intellectual property protection? The lessons of history', *Prometheus* **23**(2), 131–47.

3.4.3 Psychology and Bargaining Tools

One means of reconciling a second tier patent system like UMs may be to accept that the justification of a patent system is not so much due to it being a stimulus for invention or for disclosure, but a means for the utilisation and development of new methods and products. This in turn requires the investment of capital and the guarantees which will make such investment possible. Patent rights, in common with other intellectual property rights, exist to allow certain market actors to gain an economic *and psychological* advantage over competitors, thereby allowing the first group the opportunity to create monopolies and recoup the costs of R&D. In the past, Japanese companies have been accused of 'patent flooding' whereby large numbers of applications are made claiming minor technical improvements to existing patents: one Japanese attorney commented that it was 'psychologically difficult to work around if you have only two, even if your two are better', as compared to a Japanese company holding 40 patents in a small area.[42]

In many ways, all intellectual property rights exude an illusory effect. Indeed, as the American economist Fritz Machlup once explained, it is socially important to create an illusion on the part of innovators that imitation by their competitors will be delayed due to the strength of their patent monopoly. Under such an illusion, new investment in industrial research, development and innovation will appear promising in view of the supposedly delayed advent of competition. Whether this is true or not is irrelevant as industrial and societal progress requires both innovation and imitation; and if firms, without an adequate industrial property rights regime, anticipate rapid imitation, they will not risk expensive innovation and investment.[43] Thus, one justification of a second tier system is that it may offer the best means whereby to create this illusion which, arguably, may act as a spur to innovative and creative activity within any industrialised society.

3.5 NOT THE ONLY GAME IN TOWN: DESIGN, UNFAIR COMPETITION AND *SUI GENERIS* LAWS

When it comes to extending intellectual property to new 'things', in this case sub-patentable inventions, the options available to policymakers are to fit such

---

[42] Leith, P. (1998), 'Harmonisation of intellectual property in Europe: a case study of patent procedure', in A. Chandler (ed.), *Perspectives on Intellectual Property*, vol. 3, London: Sweet & Maxwell, p. 13.

[43] Machlup, F. (1952), *The Economics of Sellers' Competition*, pp. 555–6.

products into existing intellectual property categories or to create new intellectual property rights. In the words of William Cornish:[44]

> Intellectual property may be extended to new subject matter either by accretion or by emulation. Accretion involves re-defining an existing right so as to encompass the novel material; emulation requires the creation of a new and distinct right by analogy drawn more or less eclectically from the types already known.

UMs are an example of emulation. But emulation is inherently risky in the sense that it is essentially experimental. Accretion may be a safer option. One possibility that we explore in this chapter, may be to protect sub-patentable inventions under design law. Our research into the United Kingdom, German and Australian UM systems from a historical perspective indicates that the inception and subsequent development of second tier patent protection was a response to perceived deficiencies in both patents and designs law. Existing design legislation did not, in the German and Australian[45] experiences, protect functional innovations; whilst in the United Kingdom case, design legislation was adapted to plug the gap found to exist in the protection of minor and incremental innovations and inventions. Moreover, a consideration of the actual subject matter of protection under the various European UM laws reveals that the term 'UM' often incorporates many of the elements that would ordinarily constitute a functional or technical design. As we see from the following discussion, developing countries may have several alternatives: to introduce a United Kingdom-type unregistered design right which extends to protecting functional shapes, or a UM type of law which is specifically designed for the protection of technical functions.

Another possibility is to take advantage of unfair competition. This may be justified on the basis that, as indicated earlier, although UM laws confer exclusive property rights, the underlying rationale for having them is usually to accede to industries' call for an anti-copying right or a misappropriation tort. Indeed, the main practical justification derives from the fact that many inventions are vulnerable to unfair copying, and that in many cases, the sub-patentable ones are the most vulnerable of all.

---

44 Cornish, W.R. (1993), 'The international relations of intellectual property', *Cambridge Law Journal*, **52** (1), pp. 46 and 54–5.

45 Australian Law Reform Commission (1994), report 'Designs', discussion paper no. 58 and (1995) report no. 74.

3.5.1 Design Law

Design law in the United States

The United States design patent law is extremely demanding in comparison with other countries. Protection is available under patent law for 'any new, original and ornamental design for an article of manufacture'.[46] Furthermore, in order for a design to qualify for design patent protection, it must present an aesthetically pleasing appearance that is not dictated by function, and it must satisfy the general criteria of patentability, i.e. full novelty and non-obviousness.[47] In brief, the law does not give protection to 'new designs' or 'original designs' but rather to designs which fulfil both criteria, and requires candidates to fulfil a higher threshold of protection by requiring non-obviousness as well, a term identified more with the patent criterion of 'inventive step'.

The problems which arise stem from the impossibly high criteria imposed on designs. In practice, the rate of invalidation of design patents is high as the federal appellate courts are in the habit of routinely invalidating issued design patents that reach them in infringement actions.[48] Moreover, the thresholds of protection imposed on design patents are far higher than those applied to patents. The specific criteria of protection, and the related problems, are:

i. Novelty: a high standard of novelty is applied although the Patent Act does provide a one year novelty grace period for both utility and design patents, allowing designers the opportunity to market test their designs.[49]
ii. Ornamentality: this test can be set at high levels whereby the court must be satisfied that the design appeals to the eye as a thing of beauty.[50]
iii. Non-obviousness or 'inventiveness':[51] a major issue in respect of the last criterion is determining from whose perspective obviousness is judged. The Court of Customs and Patents Appeals *In re Nalbandian*[52] stated that,

---

[46] §§ 102, 103, 171–3, 289, US Patents Act.

[47] *Schwinn Bicycle Co. v Goodyear Tire & Rubber Co.*, 444 F.2d 295, 298 (9th Cir., 1970) (stating design criteria for obviousness covered by same language as utility patent per 35 USC 171).

[48] Denicola, R. (1983), 'Applied art and industrial design: a suggested approach to copyright in useful articles' *Minnesota Law Review*, **67**, 707; Lindgren, T. (1985), 'The sanctity of the design patent: illusion or reality?', *Oklahoma City University Law Review*, **10** (195), pp. 207, 209.

[49] 35 USC Sec. 171 (1976). The novelty requirement of patentability is expressed in 35 USC Sec. 102(a) and (b).

[50] *Bliss v Gotham Indus. Inc.*, 137 USPQ 189 (9th Cir., 1963); and generally D. Chisum, *Patent Law*, 1997, Sec. 1.04[2][c].

[51] 35 U.S.C. § 103 (1976).

[52] *In re Nalbandian* 211 U.S.P.Q. 782 (CCPA, 1981), p. 784. The CCPA re-affirmed

> in design cases, the fictitious person identified in § 103 as 'one of ordinary skill in the art' will be the designer of ordinary capability who designs articles of the type presented in the application.

The difficulties posed by the statute are exacerbated by the judicially forged and developed 'functionality' doctrine which is an attempt to prevent the circumvention of patent law by technically dictated designs. While most courts agree that design patent law should eschew fully functional designs, what remains unclear is how this is to be achieved, and the courts' usage of divergent tests makes this an uncertain area of intellectual property law.[53]

Design law in the EU

The EU design regime is governed by the Community Design Regulation 2001 and the Design Directive 1998.[54]

It may be possible to claim protection under the European design regime for minor innovations due to the broad ambit of the law. For example, the definition of 'design' is wide enough to encapsulate all types of products including parts intended to be assembled into a complex product. The only problem for minor innovations is that they may fall foul of the functional exclusion clause. Design protection will not be granted to the following subject matter:[55]

---

itself in *In re Carter*, 213 U.S.P.Q. 625 (CCPA, 1982), and *In re Rosen*, 213 U.S.P.Q. 347 (CCPA, 1982). See also *Graham v John Deere Co.* 383 U.S. 1 (S. Ct., 1966).

53 See, for example, *L.A. Gear Inc. v Thom McAn Shoe Co.*, 12 U.S.P.Q.2d 1001, 1007 (S.D.N.Y., 1989) ('The existence of … competing shoes indicates that the design configuration is ornamental, not functional'); *Petersen Manufacturing Co. v Central Purchasing Inc.*, 740 F.2d 1541 (Fed. Cir., 1984) (the court held the design of the hand-tool to be functionally dictated, but finally invalidated the design on the ground of obviousness); *Power Controls Corp. v Hybrinetics, Inc.* 231 U.S.P.Q. 774, 777–8 (Fed. Cir., 1986); *Bentley v Sunset House Distrib. Corp.*,149 U.S.P.Q. 152 (9th Cir., 1966) ('Where the "design" of a design patent is dictated primarily by functional or mechanical requirements and any pleasing aesthetic effect is only an inadvertent by-product, the design patent is invalid.'); *Avia Group International v L.A. Gear California* 7 U.S.P.Q.2d 1548 at 1553 (Fed. Cir., 1988).

54 European Parliament and Council Directive 98/71/EC of 13 October, 1998 on the legal protection of design, O.J. L289, 28.10.1998; Council Regulation (EC) No. 6/2002 of 12 December, 2001 on community designs. The 1998 Design Directive serves to harmonise the substantive aspects of member states' national laws, although no attempt has been made to harmonise the procedural aspects of the law. A major reason for the adoption of the directive is to maintain the current national design registration system, especially for the benefit of enterprises with local market concerns, in co-existence with the community design law. The Design Regulation, on the other hand, creates a unitary Community Design Right (CDR).

55 Articles 7–8, Design Directive; Articles 8–9, Design Regulation.

- features of appearance of a product which are solely dictated by its technical function
- 'interconnection' features i.e. features of appearance of a product which must necessarily be reproduced in their exact form and dimensions in order to permit the designed product to be mechanically connected to or placed in, around or against another product so that either product may perform its function. [56]

The 'technical function' exclusion is apparently to be construed narrowly – a design feature will only be excluded if the feature concerned is essential in order to achieve a particular technical result. In other words, if form follows function. Thus, functional innovation can be eligible for protection as long as one shows that the same technical function could be achieved by another different form.[57]

On the other hand, it may be easy to gain protection as there are no high thresholds of protection like inventive step; instead, one must show that the design is novel and has individual character.[58] Technically, novelty is assessed at a universal level, with the prior state of art encompassing any identical design (or a substantially similar design which differs only in immaterial details), made available to the public. Notably, a further limitation is placed on the width of the prior art taken into consideration for the purposes of novelty and individual character; in some cases, published designs will not be taken into account where the publication could not reasonably have become known in the normal course of business to specialised European business circles.[59] A design will be deemed to have individual character if the overall impression it produces on the informed user differs from the overall impression produced on such a user by any design which has been made available to the public. The same limitation as to prior art applies (i.e. must be known in the normal course of business to specialised European business circles).

The prospective rights owner in the EU has a choice as to the type of Community Design Right (CDR) required as the EU law offers a two-tier system:

---

[56] There is an exception in respect of 'interconnection' features: designs which serve the purpose of allowing multiple assembly or connection of mutually interchangeable products within a modular system will be protected. This has been taken to refer to designs destined for modular systems, and which by their nature must fit or assemble with each other. Examples of this include modular toy systems such as LEGO bricks or modular furniture systems. Article 7(3), Directive; Art. 8(3), Regulation.

[57] *Philips Electronics N.V. v Remington Consumer Products* [2002] C.M.L.R. 1329 (AGO and ECJ), para. [A34].

[58] Articles 4–8, Directive; Articles 4–7, Regulation.

[59] Article 6, Directive. The terminology used in Article 7, Regulation is almost identical.

a. the Registered CDR which confers registration-based exclusive rights
b. the Unregistered CDR which confers an automatic anti-copying right based on first marketing.[60]

Registration gives the proprietor an exclusive right of use.[61] Moreover, the registered design right offers an exclusive monopoly for a longer term of period than either the proposed EU UM right, or the patent right. The duration of protection is for an initial period of five years, but this can be renewed for four further periods of five years, up to a maximum of 25 years. Examination is not a great hurdle to pass as it is currently restricted to formalities and excluded subject matter – there is no search as to novelty or individual character.

If the proprietor opts for the unregistered CDR, a lesser 'right of copying' is conferred. The anti-copying right is further limited in that it will not be effective where use is deemed to result from 'copying the protected design if it results from an independent work of creation by a designer who may be reasonably thought not to be familiar with the design made available to the public by the holder'.[62] The unregistered CDR is a poor relation to both the registered CDR and to the United Kingdom unregistered design right (which is not affected by all these European developments, and remains distinct and *sui generis*). The main problem with the unregistered CDR is that it only lasts for a period of three years from the date on which the design was first made available to the public within the community.[63]

How popular is the design system? The following table shows a cross-comparison between the European Registration Office, the German Office and the Japanese Office.

It appears from the European Office's statistics that very little is rejected compared to the total amount of filed registrations. The next table shows that the total amount of rejected applications is less than 1 percent of received applications.

The major fault in obtaining protection for minor innovations under the EU design law is that the scope of protection may be limited to the 'overall impres-

[60] The EU design regime is governed by the European Parliament and Council Directive 98/71/EC of 13 October 1998 on the legal protection of design, O.J. L289, 28.10.1998 [Design Directive], and the Council Regulation (EC) No. 6/2002 of 12 December 2001 on Community designs [Design Regulation].

[61] Article 12, Directive; Article 19, Regulation. 'Use' is defined to cover, in particular, the making, offering, putting on the market, importing, exporting or using of a product in which the design is incorporated or to which it is applied, or stocking such a product for those purposes.

[62] Article 19(2), Regulation.

[63] Article 11, Regulation.

Table 3.7 Comparative table of design filing in 1999–2004

Year	Offices	Design registrations
1/1/2004–30/6/2004	European Office for Harmonization in the Internal Market	30,162
1/1/2004–30/6/2004	German Patent and Trademarks Office	21,078
1/1/2004–30/6/2004	Japan Patent Office	16,646
2003	Office for Harmonization in the Internal Market	19,935
2003	German Patent and Trademarks Office	54,670
2003	Japan Patent Office	31,342
From 1999 to 2002	No European Design Law	–
From 1999 to 2002	German Patent and Trademarks Office	279,584
From 1999 to 2002	Japan Patent Office	145,829

Table 3.8 Rate of refusal and grant under the community design law

	2003	2004	Total
Designs received	40,623	53,650	94,273
Refused	4	274	278
Registered and published	19,934	57,765	77,699

sion' of the design. In other words, the law will not protect an underlying function or principle but rather the appearance of the product. This is different to the United Kingdom unregistered design right which extends protection to the shape and configuration of the design, as opposed to the 'impression' that the design produces on the informed user.[64] Although the law specifically excludes protection from 'methods or principles of construction', there may be instances where the United Kingdom unregistered design right does lead to underlying technical functions and principles being protected, especially if such features are inherent in the (directly) protected shape and configuration.

[64] Article 9 Directive; Article 10, Regulation.

3.5.2 Unfair Competition

To what extent is protection offered to minor innovations within the United States and the EU under other laws?

According to Article 10bis of the Paris Convention, the following acts are prohibited on the grounds of constituting unfair competition:

- all acts of such a nature as to create confusion by any means whatever with the establishment, the goods, or the industrial or commercial activities, of a competitor;
- false allegations in the course of trade of such a nature as to discredit the establishment, the goods, or the industrial or commercial activities, of a competitor;
- indications or allegations the use of which in the course of trade is liable to mislead the public as to the nature, the manufacturing process, the characteristics, the suitability for their purpose, or the quantity, of the goods.

Many countries have a general unfair competition law which is based on fault or wrongdoing. Any infringement of an intellectual property right invariably involves fault or wrongdoing on the part of the imitator. Thus, although serving to punish the conduct of the imitator, unfair competition laws can indirectly protect proprietary rights, and in this manner can be employed to supplement the protection of an innovation.

One view as to the function of unfair competition law is that it acts as a corollary to competition law. While competition law protects the institution of competition as the chosen order of the marketplace, unfair competition theory regulates the behaviour of the various competitors with regard to their behaviour in the marketplace.[65] This is in tandem with the view that unfair competition law relates to the conduct of an imitator, rather than to what is imitated.

The second view holds that unfair competition should not be based on ethics as determining the standards of behaviour but rather should determine whether or not the result of the competitor's behaviour hinders or stifles the competitive process of differentiation and imitation.[66] This view upholds the sanctity of consumer welfare as the prime consideration, as opposed to upholding moral standards within the marketplace. For instance, imitation products on the market may reflect the dishonest business practices of the imitator, but there should be

---

[65] Kaufmann, P. (1986), *Passing Off and Misappropriation*, IIC Studies, vol. 9, Munich: VCH, p. 15.

[66] Kaufmann, P., *ibid.*

some consideration as to whether the products are also economically beneficial to the consumer if they bring the market price down and force standardisation of the product.

On the other hand, since inventions with lower inventive step do not fulfil the patent criteria, it traditionally is thought that such inventions should be left in the public domain. The problem of extending unfair competition protection to such innovations is that this move undermines the perceived need to have specific exclusions of technically necessary or functional features from copyright, design and trade mark laws. The whole rationale for these latter exclusions is that technical inventions which did not meet the patent criteria should not gain protection by the back door. Nevertheless, unfair competition laws, in some jurisdictions, will protect against such copying or imitation. Although some member states' laws have developed some jurisprudence that unfair competition laws cannot usurp other intellectual property statutory rights, especially in relation to patentable features, the law is not totally clear and there have been many instances where the tort of slavish copying has stepped in to protect such technical or functional features.

Is it now the case that intellectual property law should protect any investment or creativity which is open to imitation and copying? Have we all succumbed to Mr Justice Peterson's adage that 'what is worth copying is worth protecting'?

The danger of applying unfair competition laws liberally to protect intellectual property subject matter is that an untenable monopoly can arise in relation to a work which has either been denied protection under other intellectual property rights or ceased to be protected under any intellectual property rights.[67] Unfair competition laws can undermine the delicate balance between intellectual property laws and competition laws, thereby resulting in anti-competitive protection of minor innovations.

United States federal trade dress law

Although theoretically, functional works are excluded from federal trade dress protection, history nevertheless shows us that in practice, minor innovations can be protected under the United States federal trade dress law.

The common law of unfair trade or passing off was formally codified in the Lanham Act 1946. In light of the difficulty faced in gaining protection under the patent, copyright and state unfair competition laws for minor innovations and designs, attempts were made using the Lanham Act to obtain protection for product shapes and configurations as unregistered trade dress. The basis of

---

[67] *Massey-Ferguson v Bence Equipment & Bepco France*, TGI Versailles, 10 March, 1982 (1983) FSR 117.

protection was obtained under section 43(a) of the Lanham Act. The provision stipulates:

> Any person who, on or in connection with any goods or services, or any container for goods, uses in commerce any word, term, name, symbol, or device, or any combination thereof, or any false designation of origin, false or misleading description of fact, or false or misleading representation of fact, which … is likely to cause confusion, or to cause mistake, or to deceive as to the affiliation, connection, or association of such person with another person, or as to the origin, sponsorship, or approval of his or her goods, services, or commercial activities by another person … shall be liable in a civil action by any person who believes that he or she is likely to be damaged by such act.[68]

The notion of trade dress consists of the 'overall image used to present it to purchasers; it could thus include, to give a partial list, the product's size, shape, colour, graphics, packaging, and label'. Moreover, the definition was stretched further from the product's packaging to the *actual* product. Thus, protection is not accorded to the intrinsic inventive idea or principle, but rather to the physical nature of the product.[69]

Thus, under trade dress protection, a product could gain unlimited protection. Moreover, there appears to be little demarcation between trade dress, patent law and design law, utility patent or design patent laws. True – there is the functionality doctrine which states that functional aspects of the product will not be accorded either trade mark or trade dress protection.[70] However, a small amount of functionality will not invalidate protection if the overall effect of the product is non-functional – and this is true of most minor innovations.[71]

Functionality is only said to be present when features are essential to the use or purpose of the article.[72] In *re Morton-Norwich Products Inc.*[73] for instance, the court stated that trade dress law only excludes shapes which are *de jure* functional, rather than *de facto* functional. In this case, the court held that Morton-Norwich's shape of a spray container was not *de jure* functional because the bottle shape was not 'the best or one of a few superior designs

---

68 Lanham Act, Ch. 540, 43(a), 60 Stat. 441 (1946), current version at 15 U.S.C. 1125(a) (1988).

69 *Vaughan Mfg. Co. v Brikam Int'l, Inc.*, 814 F.2d 346 at 348, note 2 (7th Cir., 1987); *Two Pesos, Inc. v Taco Cabana, Inc.*, 112 S. Ct. 2753 at 2761 (S. Ct., 1992).

70 *Inwood Law. Inc. v Ives Lab. Inc.*, 456 US 844 (S. Ct., 1982).

71 *First Brands Corp. v Fred Meyer, Inc.*, 809 F.2d 1378, 1381 (9th Cir. 1987) (trade dress viewed as whole when determining whether dress is functional); *LeSportsac, Inc. v K Mart Corp.*, 754 F.2d 71, 76 (2nd Cir., 1985) (features of trade dress must be taken together when determining functionality).

72 *Inwood Law. Inc. v Ives Lab. Inc.* 456 US 844 (S. Ct., 1982), p. 850.

73 671 F.2d 1332 (CCPA, 1982).

available'. There was no evidence that the shape of the bottle was required to be as it was. The same function could be performed by a variety of other shapes.[74]

Unfair competition in Europe

EU member states' courts are reluctant to extend protection for subject matter which has already enjoyed protection under other intellectual property rights. The general principle is that the imitation of products is allowed if they are not protected by other intellectual property laws.[75] This is especially true in the absence of further and special circumstances, such as behaviour contrary to ethical business practices, misleading or confusing behaviour, unfair exploitation of the plaintiff's goodwill or obstructing the plaintiff from exploiting the economic benefits of his products.[76] An apt summary of the position was offered by the Dutch Supreme Court:

> Everybody should be, generally speaking, free to develop his industrial products in the best way possible and to give them the highest utility – except when exclusive rights are awarded to others on the basis of the patent or copyright statutes which prevent this – it is permitted for that purpose, in his own interest and possibly to the detriment of his competitor, to make use of the inventive ideas and results which are embodied in his product, even if (it is) only on the basis of that use confusion could be the result between his own product and the competitor's ... thus the imitation of the product of a competitor is not allowed only if, without doing harm to the function and use, another avenue could have been followed on certain points, and by not being followed confusion is created.[77]

Germany

For instance, German unfair competition law[78] prohibits all acts of competition that are unfair and capable of materially distorting competition by harming competitors, consumers or other market participants. There is no definite duration of protection as each case will depend on the facts; in general, however, the court will offer unfair competition protection sometimes to unprotected innovation for a reasonable period so as to enable the creator to exploit his competitive advantage.

---

74 *Ibid* at 1341–2.

75 WIPO, *Protection Against Unfair Competition*, WIPO: Geneva, 1994, paragraph 61.

76 See generally, Suthersanen, U. (2000), *Design Law in Europe,* London: Sweet & Maxwell, Chapter 23.

77 *Machinefabriek Thole NV v Hyster Company*, H.R., 26 June, 1953, NJ 1954, 90 at 180, in Kaufmann, p. 75.

78 Law on Unfair Competition, BGBl. I 2004, 1414 of 3 July, 2004. (Gesetz gegen unlauteren Wettbewerb – UWG).

German unfair competition law assumes that in absence of intellectual property protection, there should be nothing to impede the copying of an innovation.[79] Thus, the mere copying of a competitor's product will not necessarily be tantamount to a breach of the unfair competition law, especially where the product has been initially protected under an intellectual property right which has since expired. The courts will protect against the copying of products under the unfair competition laws only where additional elements or circumstances are present, such as:

- misleading customers and disloyalty
- causing confusion and/or unfair exploitation of the plaintiff's goodwill
- obstructing the plaintiff from exploiting the economic benefits of his products
- value of the innovation in terms of its functionality or cost-effectiveness
- short life span of the innovation
- direct misappropriation and/or regular and systematic copying.

There are exceptions. First, courts will protect innovation if in their view the plaintiffs require protection against competition in order to reap their profits, especially taking into account the expenditure of resources in innovating the new product. Second, highly original and innovative products are not generally open to copying, especially if they function as a source indicator and where the conduct of the imitator can be considered as unfair. On the other hand, if products are functional, and fail to qualify for patent and UM protection, courts are more likely to condone copying – it has always been the view that such features should be available for copying, even where there is a danger of deception since a manufacturer should not be allowed to secure a monopoly on his product, ab sent any intellectual property right.[80]

Although copying another person's product is generally allowed in the name of free trade and competition, especially where confusion is not an issue, sometimes the jurisprudence does prohibit direct misappropriation or slavish copying of another's product. The unfairness of this activity is found in the fact that in addition to the imitation, the defendant is also making direct use of the product of the plaintiff-manufacturer without any expenditure on his part, thus taking the fruits of the plaintiff's money, skill and labour, without investing any of his own.

---

[79] Rohnke, C. (1990), *Protection of External Product Features in West Germany*, 2 EIPR 41.

[80] See Suthersanen, U., *Design Law in Europe*, *op. cit.*, Chapter 23.

For direct misappropriation to be found unlawful, there are three main requirements. First, the work sought to be protected must have some characteristics worthy of protection i.e. there must be evidence of the plaintiff's investment in the work or the product has a certain distinctive character; second, the copying must have caused some actual damage to the creator of the work; finally, the imitator could reasonably have utilised other features which differed from the original product.[81]

France

In France, unfair competition (*concurrence déloyale*) law provides remedies for any civil or legal wrong based on the finding of 'fault'.[82] The most frequently sought circumstances under which an action on unfair competition is brought, concerning an intellectual property subject matter, are confusion, rival disparagement or unlawful attacks. However, it is debatable as to whether imitation, *per se*, is insufficient to constitute an action, especially where the work or product is not protected by any intellectual property right. As a fundamental principle, in the absence of any intellectual property rights, the product can be freely appropriated by all. However, such liberty is not without its limits, especially where there is excessive usage to the detriment of commerce. Thus, the act of imitation can develop into a tortious act if the reproduction was for an unfair purpose or if the resulting confusion between products cannot be justified on other grounds.

The general principles governing imitation of products can be reduced to the following principles:

- Copying an item in the public domain cannot amount to fault.
- Unfair competition law cannot be invoked as it is not a civil wrong to reproduce an unprotected object even if damage is caused – a further element is required.
- The reproduction of products belonging to the public domain is wrong if this is done with a view to creating a confusion in the public concerning the origin of products sold.

---

[81] See *Lego Bricks*, GRUR, 1964, p.621; and *Tchibo/Rolex*, BGH, GRUR 1985, p.876.

[82] The law of unfair competition is not governed by special texts but is part of the civil law as governed by Article 1382 of the French Civil Code. Some actions are also based on Article 1383 Civil Code which covers acts of negligence. In general, there are two main areas of unfair competition which concern intellectual property rights: *concurrence déloyale* (which addresses issues of interference with and obstruction of competition) and *concurrence parasitaire* (which deals with the usurpation of the plaintiff's distinctive achievement to the defendant's benefit). See Suthersanen, *Design Law in Europe*, *op. cit.*, Chapter 23.

- The taking of fruits of another's industry and investment is unjust enrichment and parasitic competition (*concurrence parasitaire*).[83]

3.5.3 Specific Responses to Specific Technologies – *Sui Generis* Laws

Finally, there is the option of protecting innovations under a *sui generis* law i.e. the emulation option. This strategy is sometimes chosen to enable innovators in certain fields of science, technology or business to appropriate the outputs of their research in a more effective and balanced manner that, policymakers believe, would have been difficult to achieve under existing regimes. Examples include plant variety protection and semiconductor chip design protection.

In respect of the semiconductor technology, a modified copyright approach was adopted in the 1984 United States Semiconductor Protection Chip Act. This legislation is historically interesting as it re-introduced an approach which had increasingly fallen out of favour in the protection of scientific innovation: a non-registration system based on criteria other than novelty or inventiveness. This legal formula was subsequently adopted by the United Kingdom in its 1988 unregistered design rights regime, but extended to all types of functional designs. Intriguingly, the same formula was incorporated into the EU Unregistered Community Design Right (see page 72) protection.

Another type of industry-specific law is the United States Vessel Hull Design Protection Act (incorporated as 17 U.S.C. Chapter 12) which is the result of a rather bizarre lobbying endeavour by the boat industry. Protection is granted to an original design of a vessel hull, vessel plug or vessel mould which makes the vessel attractive or distinctive in appearance to the purchasing or using public.[84] Originality, due to the industrial basis of the right, is even equated to prior art and is defined thus: 'the result of the designer's creative endeavour that provides a distinguishable variation over prior work, pertaining to similar articles which is more than merely trivial and has not been copied from another source'.[85]

The no-registration/low threshold formula is, of course, a variation on the copyright approach. Many policymakers tend to equate copyright law with 'art' and 'music' whereas in reality, the copyright approach is extremely attractive to short-lived technologies and industries such as the toy, fashion and textile industries which are fast moving, quickly imitated and in need of immediate and automatic protection, without the encumbrance of application or registration

---

[83] Suthersanen, U., *Design Law in Europe*, *op. cit.*, Chapter 23; Kamperman Sanders, A. (1992), *Unfair Competition Law – The Protection of Intellectual and Industrial Creativity*, Oxford: Clarendon Press, p. 29.

[84] § 1201(a).

[85] Excluded are staple, commonplace and designs dictated solely by a utilitarian function of the article that embodies it.

costs. The lower threshold (based on interpretations of terms like 'commonplace' and 'originality') is advantageous for industries which customarily rely on the prior state of art and which represent incremental, rather than massive, design improvements. This lower threshold also allows industries to embark on market testing for their products without any loss of protection. Moreover, copyright laws gravitate towards the individual innovator, creator or designer and are thus more friendly to either the sole inventor or an SME-type inventor.

3.5.4 Interaction of Utility Model, Design and Unfair Competition Laws

In essence, the choices facing policymakers in respect of sub-patentable inventions are far from straightforward. They may choose to leave them unprotected, to lower the inventive step threshold so that they become patentable, or they may seek to create alternative legal means of protection. One alternative, as we have seen, is to introduce a liability rule-based system such as a statute-based tort or misappropriation or unfair competition law.

Since the main practical justification for UM derives from the fact that many inventions are vulnerable to unfair copying, and that in many cases, the sub-patentable ones are the most vulnerable of all, in the alternative (or cumulatively) one can also introduce a hybrid property rights system such as industrial design rights and/or utility models. Ironically, then, the classic UM regime, as exemplified by German legislation, was originally conceived as a form of design protection.[86]

These systems could operate in place of patents or copyright, as the case may be, or could alternatively operate in parallel, so that patents and copyright could then be used only for those inventions and creative works displaying a demonstrably high inventive step or level of originality. Some countries such as Germany, Japan and China employ all these alternatives, and the local innovator has a choice of protection ranging from patent to design to utility model to unfair competition laws.

The fundamental reason for such a cross-over of rights is that intellectual property protection of three-dimensional products cuts across the traditional intellectual property demarcation lines. The interface, for example, which occurs between design law and patent law is actually reflected in UM law. A consideration of the actual subject matter of protection under various laws reveals that

[86] Krasser, R. (1995), 'Developments in utility model law', *International Review of Industrial Property and Copyright*, 950, 953; Ladas, S.P. (1975), *Patents, Trademarks, and Related Rights: National and International Protection* (551), Cambridge, MA: Harvard University Press, p. 952; Janis, Mark D. (1999), 'Second tier patent protection', *Harvard International Law Journal*, **151**, 159.

the term 'utility model', in many cases, incorporates many of the elements that would ordinarily constitute a functional or technical design since most UM laws limit protection to objects with spatial form. Indeed, in countries such as Italy and Portugal, the 'three-dimensional form requirement is so strong that careful interpretation is needed before it can be decided that it is the invention which is protected'.[87]

3.6 STICKING WITH PATENTS: THE CASE OF THE UNITED STATES

The United States is, perhaps, the world's biggest intellectual property producer. Nevertheless, the United States has never had second tier patent protection, though there are some calls for such a system to exist.[88] How far, then, does the United States intellectual property law protect minor innovations? There is limited scope for protecting innovation under the United States design patent and trade dress laws; however, highly functional innovations usually claim protection under patent law. One can, of course, speculate on the reasons why the United States has felt no need for a second tier regime. Perhaps as a net exporter of intellectual property goods, the United States has no need to worry about reducing the level of imports – the fact that it is a major exporter of intellectual property goods, including patented goods, shows that the current system of patent law and design patent law works for this particular economic and social environment. Another possibility is that most of the innovations and inventions do not emanate from SMEs but rather from large transnational corporations, and the level of competition amongst these firms is high enough to maintain the incentive and momentum to create – no further legislative activity is required to furnish additional inducement. Finally, one can also state that the main concern of the United States legislators is not so much its domestic market, but protecting its intellectual property goods outside its jurisdiction, and its national

[87] Italian Legislative Provisions on Industrial Model Patents (Royal Decree No. 1411 of 25 August, 1940, as amended in 1987; the reverse is true in countries which do not have utility model laws such as the United Kingdom where secondary technical matter can be protected under the unregistered design right, Cornish, W.R. and D. Llewelyn (2003), *Intellectual Property*, 5th edn, Sweet & Maxwell, at para. 3–32.

[88] Prior Commercial Use and Patent Infringement: Hearings Before the House Committee on the Judiciary, Subcommittee on Courts, and Intellectual Property, 104th Cong. 21 (1995) (statement of Professor Karl F. Jorda), cited in Janis, Mark D. 'Second Tier Patent Protection', *Harvard International Law Journal*, **40** (151), 155. Interestingly, in North America, Mexico is the only country to have introduced a second tier regime in the early 1990s. See Pemberton, G.A. and Mariano Soni Jr. (1992), 'Mexico's 1991 Industrial Property Law', *Cornell International Law Journal*, **25** (103), 123.

policy has focused on maintaining a high intellectual property regime outside its borders.

But however tempting these conclusions are, they do not hold up against recent studies which show, first, that small to medium sized corporations dominate the patenting scene, and second, that the rate of grant of patents is high compared to the low rate of invalidation.[89] Part of the reason stems from the 1982 patent law reforms which opened the path to higher patenting levels in the United States.[90] Moreover, developing countries must note that part of the innovative environment and climate is engendered by the United States government which has, since the 1980s, actively pursued a patent policy which has promoted the interests of patent owners.[91]

3.6.1 Summary of Analyses

Kortum and Lerner analysis

From the mid 1980s onwards, the United States Patent and Trademark Office (USPTO) has been besieged by an enormous increase in the number of patents filed. This phenomenon was investigated in the late 1990s by Kortum and Lerner.[92] Three main hypotheses were offered. The first hypothesis was that perhaps the United States is a favourable destination for patent applications due to recent changes of the laws which have broadened the rights of patentees greatly. However, the report noted that the recent surge in domestic patenting was particular to the United States only. A second hypothesis was that the increase in patenting was a reflection of the fertile technological environment within the United States which had originated from the following factors: new firm formation and innovation in high technology sectors; the tremendous growth of financial intermediaries funding start-ups from the mid 1980s onwards; and the shift in R&D towards more applied activities, thus raising the yield of patentable discoveries. Nevertheless, the report noted that from the

---

89 Kortum, S. and J. Lerner (1997), 'Stronger protection or technological revolution: what is behind the recent surge in patenting?', National Bureau of Economic Research working paper 6204, September.

90 Jaffe, A.B. and J. Lerner (2004), *Innovation and its Discontents: How our Broken Patent System is Endangering Innovation and Progress, and What to do About it*, Princeton, NJ: Princeton University Press; and Katz, M.L. and J.A. Ordover (1990), 'R&D competition and cooperation', *Brookings Papers on Economic Activity: Microeconomics* (137), 92.

91 Mowery, D.C. and N. Rosenberg (1993), 'The US national innovation system', in Richard R. Nelson (ed.), *National Innovation Systems: A Comparative Analysis*, Oxford: Oxford University Press, p. 59.

92 Kortum, S. and J. Lerner (1997), *op. cit.*, National Bureau of Economic Research, Working Paper 6204, September.

empirical data, the increase in patenting activity was being experienced across the whole spectrum of different technologies. The third hypothesis is the most interesting as it has also been advocated by other analysts as to why there has been an increase in patent grants in the United States. There may be some regulatory capture of the USPTO and other relevant branches of government, whereby entrenched incumbents well versed in the usage of intellectual property rules and institutions use their knowledge in influencing policy, legal and administrative changes to enhance their position relative to new entrants. In the end, Kortum and Lerner do suggest that the patenting surge was driven by changes in the management of R&D activities. Moreover, new and less established patentees were more aggressively exploiting the patent system in the late 1980s as compared to previous periods. The counterargument is that the research investment climate in the United States has generally flattened from the 1990s onwards – the study suggested that although the amount of United States domestic patent applications granted by the USPTO declined very little, the intensity of research effort had not necessarily risen during the same period.

Jaffe and Lerner analysis

In a more recent study, Jaffe and Lerner suggest a similar pattern.[93] They point to the 1982 patent law reforms which opened the path to higher patenting levels in the United States. The creation of the Court of Appeals for the Federal Circuit has strengthened the protection granted to patent holders by substantially increasing the rate of patent validity. Since its inception, the court has upheld approximately 80 percent of the cases argued before it, which is an increase from the pre-1982 rate of 30 percent for the federal branch.[94]

Antitrust factors – the Mowery-Rosenberg analysis

A similar conclusion is also reached by the Mowery-Rosenberg analysis of the United States National Innovation System which shows that the United States government has, since the 1980s, actively pursued a patent policy which has promoted the interests of patent owners.[95] These authors also argue that the pro-patent environment was aided by a more relaxed antitrust policy whereby federal antitrust suits against high-technology firms were dropped, and more

[93] Jaffe, A.B. and J. Lerner (2004), *Innovation and its Discontents: How our Broken Patent System is Endangering Innovation and Progress, and What to do About it*, Princeton, NJ: Princeton University Press.

[94] Katz, M.L. and J.A. Ordover (1990), 'R&D competition and cooperation', *Brookings Papers on Economic Activity: Microeconomics* (137), 92.

[95] Mowery, D.C. and N. Rosenberg (1993), 'The US national innovation system', in Richard R. Nelson (ed.), *National Innovation Systems: A Comparative Analysis*, Oxford: Oxford University Press, p. 59.

domestic mergers and acquisitions were allowed on the ground that international competition significantly reduced the dangers of domestic market power.

Regulatory capture – the Lemley analysis

The notion of regulatory capture and institutional culture is a key theme in the analysis undertaken primarily by Lemley (with others at various times). The first and foremost reason as to why there is an increase in United States domestic patenting, these studies suggest, is the failure of the USPTO to stem 'bad' patents or to indeed cope effectively with the flood of patent applications. The inability of the examiners, and at times, their disinclination, to reject applications has led in the past to the United States being the first country to grant software and Internet 'business method' patents – two highly controversial areas.[96] The reasons for this, Lemley suggests, are the strong structural and psychological pressures on examiners to issue patents rather than reject applications, no matter how weak the alleged invention seems. Important factors explaining this pressure include:

- an inability to spend the necessary amount of time examining each application – each application requires time spent reading the application, reading the submitted prior art, searching for and reading prior art in databases, comparing that prior art to the application, writing a memorandum, reading and responding to the response to this action, iterating the last two steps at least one time and often more, conducting an interview with the applicant, and ensuring that the diagrams and claims are in form for allowance;[97]
- the impossibility of coping with the rather bizarre United States prosecution rules which make it virtually impossible to reject a patent once and for all – thus after 'final rejection', an applicant can still re-file the same application – called a 'continuation' – an unlimited number of times;[98]
- the implicit incentive within the system to award patents – since examiners must give written decisions for rejecting an application but not for allowing one, this gives them more incentives to allow rather than reject an application.[99]

---

[96] Lerner, J. (2000), 'Where does state street lead?: a first look at finance patents, 1971–2000', May, accessed at http://ssrn.com/abstract=224895.

[97] See also Merges, R.P. (1999), 'As many as six impossible patents before breakfast: property rights for business concepts and patent system reform', *Berkeley Technical Law Journal*, **14** (577), 609.

[98] 37 C.F.R. §§ 1.60(b)(4), 1.62 (2000).

[99] Thomas, J.R. (2001), 'Collusion and collective action in the patent system: a proposal for patent bounties', *University of Illinois Law Review*, **1** (305).

Lemley reasons that the only way for an examiner to guarantee that an application is finally disposed of is to issue a patent. Examiners who want credit for dispositions therefore have a strong incentive to issue patents to persistent applicants, rather than to continue rejecting the applications. He is supported in this conclusion by others.[100]

3.6.2 Does the United States Need a Second Tier System?

The United States patent law is unique in the sense that it is the only country retaining a first-to-invent, as opposed to first-to-file system. There are no explicitly defined subject matter exceptions such as lists of what is not patentable as in European patent law. The criteria for patentability are novelty, utility and non-obviousness. In practice the latter two are not very different from industrial applicability and inventive step as in numerous other jurisdictions. One interesting feature is that unlike in Europe, inventors or their successors in title who disclose their invention are allowed a grace period of up to 12 months during which they can still file a patent despite this prior disclosure.

The United States has and is still reviewing the possibility of introducing second tier patent systems to address several problems and issues in their current patent system. Advantages of a lower tier patent protection being envisaged include the following:

- to raise fees and reduce applications for top tier patents. This would provide efficiency, with the patent office having to examine fewer standard patent cases;
- to eliminate the coverage gap between designs and UMs;
- to quickly and cheaply protect inventions of incremental nature and having short commercial life in a fast moving technology market as is the United States;
- to discourage defensive patenting by firms as standard patents become much more expensive.

One view is that the United States, in not introducing a second tier protection regime, together with its equally strict design patent regime, has an industrial policy that emphasises free competition within the product and innovation markets. Developing countries wishing to emulate the innovative spirit and culture of this major industrial country may conclude that the absence of a se-

---

[100] See Doern, G.B. (1999), *Global Change and Intellectual Property Agencies*, Pinter: London and New York, p. 18 agreeing with the rest that the USPTO's organisational structure is now based on a 'process and customer-based' structure.

cond tier regime is a contributing factor. However, whilst some may be led to think that this is the best course for a pro-competitive innovative climate, it can be argued that the United States intellectual property system is difficult to transplant to other jurisdictions.[101]

First, it is difficult to appreciate the scope of its patent law without recognising how this interacts with other laws such as trade dress and competition laws. Thus, one should note that in some cases of unfair copying of innovation in the United States, firms do resort to unfair competition laws.[102] Second, the legal and political infrastructure of the United States (as ensured by the constitution) ensures that there is a balance between intellectual property rights owners and the public interest, albeit that this balance has shifted in recent years towards a pro-intellectual property position.[103] Nevertheless, the legal infrastructure within the United States ensures that a check is made at some point – either by Congress or by the courts. Developing countries may not be able to implement such checks and balances.

Conversely it can be argued that the United States has no need for a second tier system as the current patent system requirements are relaxed enough to protect minor innovations including by small and medium sized firms. First, the USPTO and the courts have been granting patents to inventions which show a low level of inventiveness.[104] Second, analyses undertaken by Lerner et al. since 1997 to 2003 suggest that the increase in patenting activity in the United States may be due to increased innovative activities of smaller firms. The studies show that whilst there was no evidence to suggest that larger firms were particularly innovative, there was nevertheless disproportionate representation of smaller firms, especially those with stronger academic ties.[105] Finally, the Lemley and

---

[101] F. Abbot (2006), 'Intellectual property provisions of bilateral and regional trade agreements of the United States in light of U.S. federal law', ICTD/UNCTAD issue paper no. 12.

[102] However, there are limitations as the courts are not particularly keen to extend unfair competition protection to highly utilitarian, functional or technical designs and products, and never processes. *Sears, Roebuck & Co. v Stiffel Co.*, 376 US 225 (S. Ct., 1964); *Compco Corp. v Day-Brite Lighting, Inc.*, 376 US 234 (S. Ct., 1964); and *Bonito Boats v Thunder Craft Boats, Inc.*, 9 USPQ 2d 1847 (S. Ct., 1989); R. Horta, 'Without secondary meaning, do product design trade dress protections function as infinite patents?' (1993) *Suffolk University Law Review* 113; *Ferrari s.p.a. Esercizio Fabriche Automobili v Roberts* 20 USPQ2d 1001 (6th Cir. 1991); but *cf Traffix Devices, Inc. v Mktg Displays, Inc.*, No. 99–1571 (S. Ct., 20 March, 2001).

[103] See, for example, the balance between free speech, as ensured under the First Amendment, and copyright law, as ensured under Article 1, Clause 8, US Constitution as discussed in *SunTrust Bank v Houghton Mifflin Co.*, 252 F.3d 1165 (11th Cir. 2001).

[104] See United States Federal Trade Commission (FTC) 'Report 2003', pp. 9–12.

[105] Kortum, S. and J. Lerner (1997), 'Stronger protection or technological revolution: what is behind the recent surge in patenting', National Bureau of Economic Research

Thomas studies show that the USPTO, sometimes following important court decisions, has boldly extended patent protection to new technologies such as biotechnology, software and Internet business methods without being unduly hampered by lack of documentation relating to the state of the art or the lack of training of its patent examiners in these new technologies.[106]

Indeed, one can be bold and assert that the *lack* of a second tier system puts pressure on the patent system to accommodate minor inventions. Indeed, the 2003 FTC Report has expressed concern in the United States as to the maintenance of the proper balance between patent protection and competition, as the current swing is too much in favour of patents. Is it arguable that the introduction of a second tier system in the United States would contribute to keeping the public domain freer than under the more expansive patent system with its longer terms of protection?

working paper 6204, September; Jaffe, A.B. and J. Lerner (2004), *Innovation and its Discontents: How our Broken Patent System is Endangering Innovation and Progress, and What to do About it*, Princeton, NJ: Princeton University Press.

106 Lemley, M. (2001), 'Rational ignorance at the patent office', *Northwestern University Law Review*, **95** (4); Lerner, Joshua (2000), 'Where does State Street lead?: A first look at finance patents, 1971–2000', May, accessed at http://ssrn.com/abstract=224895; Merges, R. (1999), 'As many as six impossible patents before breakfast: property rights for business concepts and patent system reform', *Berkeley Technical Law Journal*, **14** (577), 609; and also Thomas, J.R. (2001), 'Collusion and collective action in the patent system: a proposal for patent bounties', *University of Illinois Law Review*, **1**, 305.

4. Policy considerations for governments

Uma Suthersanen and Graham Dutfield

In essence, the choices facing policymakers in respect of sub-patentable inventions are far from straightforward. Questions for governments are:

- Should we leave sub-patentable inventions unprotected (sub-patentable meaning those inventions which show little or no inventiveness)?
- Should we lower the inventive step threshold under the standard patent law so that more inventions, including minor sub-patentable innovations, become patentable?
- Should we seek to create alternative legal means of protection such as a tort or misappropriation law, or a hybrid property rights system such as design rights?

The fundamental precondition for a decision on whether or not to adopt UM protection is that developing countries must map out and evaluate their own industrial and innovation base, and identify its current and long term economic requirements. The following points elaborate what type of specific considerations and questions need to be asked.

4.1 THE RIGHT TYPE OF 'INTELLECTUAL PROPERTY INSTITUTIONAL ORDER'?

Intellectual property systems are more than just pieces of legislation, and may best be viewed as public policy regulatory institutions. As such, they consist of the relevant statutes, rules and regulations plus the government agencies, courts and professional people involved in interpretation, implementation, enforcement and reform. Institutions are not static but evolve over time, and they operate in different ways according to the context. Thus, they may be appropriate and functional in one context but inappropriate and dysfunctional in another, and it can be very difficult to predict how well such an institution transplanted to a completely different cultural, political or economic milieu will work in practice. History can offer some pointers, or can at least help us to reflect on the likely preconditions for a successful transplantation.

Take the case of Germany and the United Kingdom, whose historical experiences are quite different. In the former country there was no coherent or unified industrial property policy until the creation of the German free trade area in the late nineteenth century. The first patent law in the German Reich was introduced relatively late in 1877 against a background of anti-monopoly and anti-patent movements. Part of the reason for this push for industrial property protection was the change in the economic landscape of Germany. German territories, especially Prussian ones, were agrarian at the beginning of the nineteenth century. Between 1850 and 1870, however, Germany was industrialising rapidly, and industrial leaders such as the Siemens firm organised pro-patent lobbies. UM law was first introduced in 1891 because it was thought that patent law was not suitable for all types of inventions. The important factor to remember is that the German local market is used to the UM system and there is knowledge about it.

Importing such a system into any developing, or indeed, developed country would require educating the local industrial users as well as the patent community. This is clear from the historical experience of the UM system in the United Kingdom. The UM system in the United Kingdom faltered as it did not have the support of the main lobbying force in relation to patent reform. Nineteenth century patent agents were not keen to reform the existing patent regime; neither, however, were they keen to introduce a new registration system which may have undermined their existing stronghold under the patent law.

The key question is:

- Does the government have resources to create the right type of 'intellectual property institutional order' which will support the UM systems?

4.2 IMPORTER OR EXPORTER OF INTELLECTUAL PROPERTY PRODUCTS?

There is an accepted view within some industrialised nations that sub-patentable inventions should be protected, and a second tier patent regime is therefore necessary. This view is being increasingly accepted within other nations too, especially those countries which are heavy intellectual property importers as opposed to exporters and where patent law's perceived inability often to cope with an innovative environment dominated by *local* incremental inventions may be seen as constituting a major deficiency. Consequently, much of the rationale in pursuing a second tier patent law is grounded in the fact that the second tier system may respond to or avoid the shortcomings of the patent law in fostering local innovation.

The 2004 OECD study by Park and Lippoldt on international licensing and intellectual property rights indicates that the strengthening of intellectual property rights has had a net positive effect on technology transfer via licensing during the 1990s. This data was based on licensing activities of United States multinationals (as opposed to SMEs) and also on international licensing alliances between firms in developing and developed nations. The bulk of licensing income is apparently derived from developed countries. However, the study also indicted that among the emerging market economies, the leading participant licensee nations are the Asian countries i.e. Singapore, Korea, Taiwan, Hong Kong and China. The OECD report concluded that securing licensing contracts with developed country partners such as the United States, United Kingdom and Japan appears to be an important strategy in the developmental catch up process. However, two factors should be mentioned. Technology transfer comes at a price: the recipient country pays for such transfers, and the recipient country must have a dedicated and efficient intellectual property rights structure.[1]

The key questions are:

- Is the country a net importer or exporter of IP goods?
- Is there a need to limit imports of foreign goods by encouraging the local industrial base?

4.3 COPYING: CREATIVE IMITATION OR MISAPPROPRIATION?

The underlying rationale for UM rights, in practical and policy terms in many countries, is that these rights are there due to industries' lobbying for an anti-copying right or a misappropriation tort rather than as a spur to local innovation. Indeed, the main practical justification derives from the fact that many inventions are vulnerable to unfair copying and that in many cases the sub-patentable ones are the most vulnerable of all. If one accepts the 'unfair copying' argument favouring IP protection, it follows that any subject matter evincing some sort of intellectual or capital investment, and which is open to imitation and copying, should be considered a worthy intellectual property good requiring protection. Indeed, some products, once invented and marketed, are especially susceptible to being copied at negligible or zero cost. Often, these products come from industries whose innovations tend to be small, incremental and cumulative in

[1] Park, W. and D. Lippoldt (2004), 'International licensing and the strengthening of intellectual property rights in developing countries', OECD trade policy working paper no. 10, December.

character. Evidently, it is for these industries that a UM system is likely to be most beneficial.

However, whilst some industries tend to be very enthusiastic about low cost, fast protection regimes (such as a no-examination UM system or a no-registration property right), other industries are highly suspicious of such systems, especially when they are viewed as curtailing industries' right to innovate on the basis of 'creative imitation' and access to a large public domain. So one really needs to know to what extent copying is a problem in the different industries, and whether such inventions are better left in the public domain and open to imitation, adaptation and incremental improvement.

As one eloquent defender of the public domain expressed it, albeit in the context of copyright, 'information is a vast lake. All of us dip into it, and millions of us contribute to it by the tankerful or the teaspoonful. If we try to account for every drop, and fight over the ownership of this or that gallon, all but the lawyers will drown in the flood.'[2] The point being made is not just that enclosing the 'intellectual commons' with property right claims leads to excessive litigation and legal uncertainty, but that innovation may be hindered as a consequence of the reduced access to information. Indeed, the importance of the public domain is often overlooked. This is unfortunate because a strong public domain is necessary for the IPR regime to perform its public policy function. According to Litman, 'the public domain should not be understood as the realm of the material that is undeserving of protection, but as a device that permits the rest of the system to work by leaving the raw material of authorship available for authors to use.'[3]

The key questions are:

- To what extent is copying a problem in the country, and which types of industries are facing this problem?
- Are such inventions better left in the public domain and open to imitation, adaptation and incremental improvement?
- Is there an economic argument as to why such inventions should be protected (e.g. the importance of maintaining or encouraging or developing a particular industry)?

Note that 'inventions' are sometimes looked upon by courts and industries as 'three-dimensional products.'

---

2 McCue, J. (1995), 'Who cares about Dickens' heirs?', *The Spectator* 25 February, p. 24.

3 Litman, J. (1990), 'The public domain', *Emory Law Journal* **39**(4), 965–1023, p. 968.

Further queries are:

- If there is a need for a quick, cheap and efficient protection regime, should it be a non-registration or registration system?
- Are there already alternative forms of protection in the country which are being offered to such inventions such as trade mark laws, unregistered design rights, unfair competition and registered design laws?

4.4 NATURE OF INNOVATOR AND INNOVATION

It is often claimed that UM systems are particularly advantageous for SMEs, especially in developing countries. For one thing, it is quite likely that SMEs have a large presence in those industries where cumulative innovation is the norm and unfair copying is rife. Indeed, it is often argued that a cheap and rapid second tier patent regime would improve the legal environment for SMEs, especially those which are engaged in an ongoing process of innovation and adaptation. This is more so in relation to certain product sectors which are concerned, not so much with revolutionary technological breakthroughs, but more so with incremental or improvement innovation. For example, one reason for the draft European Commission Directive is the perceived need for a rapid and cheap protective regime for such minor innovations in the following industries: toy manufacturing, clock and watchmaking, optics, microtechnology and micromechanics.

For another, it may even be that more innovations, both of the breakthrough and incremental varieties, emanate from SMEs than from larger multinational conglomerates. If this is so, it is important to gauge whether the current patent regime is attuned to the needs of SMEs and the types of inventions they produce. Many inventions which originate in SMEs have a lower standard of inventiveness, and are prime candidates for free-riding activities by competitors. Consequently, UMs may be highly pro-innovation and consequently good for the national economy. Another reason why a second tier patent system may be good for SMEs is that the cost factor may inhibit them from using the standard patent system as much as they would desire. The second tier patent regime is viewed as the ideal solution as it is a system geared towards the needs of SMEs, including in terms of cost.

The key questions are:

- What is the percentage of SMEs within the region?
- How much innovation, which is important to that region's economy, emanates from these SMEs?
- What type of innovation is being produced?

- Is the current patent regime suited to their needs in terms of criteria of protection, costs and ease of use?

4.5 HISTORICAL AND ECONOMIC CLIMATE OF A COUNTRY

Sometimes the decision whether to have UMs can only be made if the following questions are answered:

- What sort of development and innovation is being encouraged by the government?
- If the country's strength lies in the chemical or pharmaceutical industries, is there really a need for a UM system?

Alternatively, if the strength lies in minor innovation in engineering or optical technologies, will the UM system spur or maintain economic growth?

4.6 THE FINAL OPTIONS

There are three options which a developing country can consider:

- Status quo approach – a developing country can accept the existing intellectual property regime, without introducing any new right.
- Accretion approach – a developing country can adjust the existing intellectual property regime without introducing a UM right. This can be done by extending existing intellectual property rights to new subject matter (such as sub-patentable or functional innovation) by re-defining an existing right to encompass the new material.
- Emulation approach – emulation involves creating new hybrid rights. If a developing country does not have such a right, it would be the most expensive option in the short run. However, this expense is an immediate real cost which may be offset by long term benefits to the industrial environment such as increased international licensing opportunities.

PART TWO

Harnessing minor innovation: national studies

5. Singapore

Kit Boey Chow, Kah Mun Leo, Susanna Leong

A very detailed study of Singapore was undertaken under the aegis of the Singapore Intellectual Property Academy. This chapter reflects that study which provided empirical evidence in order to assess whether Singapore should adopt a UM regime.

5.1 METHODOLOGY

Three empirically based approaches are adopted in order to offer policymakers in Singapore a thorough understanding of the specifics of the Singaporean economy and of the needs and interests of business in Singapore.

The National Survey

The first approach was to survey innovation throughout the Singapore economy and to find out how much innovation is going on, by whom and in what areas. This survey indicated the following breakdown of research and development (R&D) companies in Singapore:

- 40.1 percent of them are foreign companies (that is, wholly owned foreign companies as well as those with more than 30 percent foreign ownership)
- 48.7 percent are local small and medium sized enterprises
- 11.2 percent are local large enterprises.

Innovation was measured largely by R&D expenditure and patent registration figures. It was found that the annual compounded R&D growth rate from 1994 to 2002 has been 14.2 percent, and that patent applications by the chosen companies had grown between the same years from 263 to 936. While these findings strongly suggest that Singapore is becoming increasingly innovative, one should be careful about drawing too firm conclusions. First, while R&D expenditure increases represent a growing commitment to innovation, it cannot be assumed that greater R&D input necessarily leads to greater innovative output. Second, patent statistics can hardly be treated as accurate quantifiers of innovation levels. For all we know, upsurges in patenting activity may have much to do with institutional changes or with different business strategies.

The Targeted Survey

The second approach was a targeted survey covering 104 firms from both manufacturing and services sectors with the aim of assessing their levels of innovation and identifying their views relating to the feasibility of introducing a second tier patent system. The survey was based on the results of questionnaires and face to face interviews, and sought to determine the following:

- the profile and demographics of respondents
- levels of innovation and their impact on revenue
- awareness and usage of intellectual property, locally and abroad, as well as the types of intellectual property systems used
- views on protection of minor innovations and the viability of UMs in protecting them.

The level of innovation within them was moderately innovative with slightly more than one-third having 25 percent or more revenue generated in the past three years from new improved products and services. A majority felt the need for minor inventions to be protected against imitation through a cheaper and more rapid form of protection of innovation. More than half of them felt that Singapore would be disadvantaged without a UM system though opinion was quite evenly divided among them on using it. This might be due to the lack of information on issues like pricing and cost relative to standard patents, protection scope and duration coverage. Such information is necessary in a decision making process given that these issues vary between countries in accordance with their specific economic and industrial needs.

In addition, moderate innovation was found in our survey of 104 firms as slightly more than one-third (35.1 percent) of the respondents had 25 percent or more revenue generated from new improved products and services over the past three years. Intellectual property as a strategic tool is still in its infant stage as less than half of them made use of either local or foreign intellectual property systems. The most widely used local and foreign intellectual property were registered trade marks and patents respectively. A majority agreed on the protection of minor innovations and the necessity for a cheaper and more rapid form of protection of innovation to promote industrial growth. More than half of them felt Singapore would be disadvantaged without a UM system though opinion was quite evenly divided among them on using it should it be implemented.

It was clear that the Singaporean industry does not really experience an unfair copying problem. Of the respondents 60.8 percent reported that no copying had occurred to any of their minor services or products. Moreover, 84.8 percent of the respondents were overwhelmingly in support of a large and free public domain to assist follow-on innovation.

The Case Studies

The third approach was to present case studies of three innovative firms operating in different industrial sectors based on face to face interviews. The idea was to gain deeper insights into the factors affecting innovation activities, the company's strategy and philosophy with respect to innovation, and the top executive's views on UMs for Singapore. The executives interviewed felt the intellectual property regime in Singapore is robust and of world standard. However, awareness was relatively low compared to other developed countries. The use of intellectual property was mainly to protect core competencies and prevent competitors in illegitimate copying of their products and designs, and also to signal potential clients of the uniqueness of their products to be used as bargaining chips at the negotiation table. It was also considered necessary to encourage innovation in Singapore by a cheaper and faster IP-protection system.

5.2 INNOVATION IN SINGAPORE

In a global business environment driven by intense competition from regional competitors, the trend in Singapore's innovation and R&D acts as a proxy, though an imperfect one, that might intuitively explain the rise in the number of patent applications and awards. Under the annual R&D survey for the year 2002, 519 valid responses were obtained.[1] The ownership distribution of the surveyed firms is detailed in Figure 5.1.

In the period 1994–2002, total R&D expenditure increased 2.9 times from S$1,175 million to S$3,404.7 million (Table 5.1). This represented an annual compounded growth rate of 14.2 percent during the eight year period. The private sector was instrumental in propelling the high growth as it accounted for more than 60 percent of total expenditure each year while the remaining 40 percent were from the government, public research institutes and the higher education sector.

1 Agency for Science, Technology and Research, Singapore.

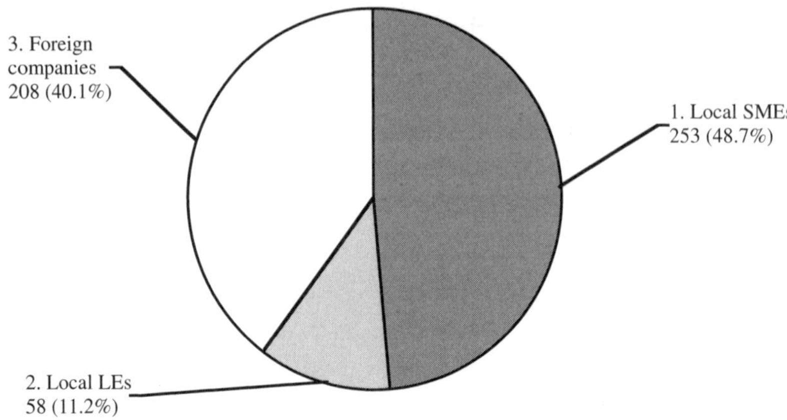

Figure 5.1 R&D survey respondents in Singapore by ownership, 2002[2,3]

5.2.1 Patent Applications

Between 1994 and 2002, the number of patent applications by the respondents rose from 263 to 936 or at an annual compounded rate of 17.2 percent (Figure 5.3). Patents awarded grew from 58 to 451 at 29.2 percent per annum. The proportion of awards to applications also increased from 22 percent in 1994 to 48 percent in 2002, reflecting proportionately more awards per 100 applications.

A total of 936 patents were applied for in 2002. The private sector accounted for 660 patents or 70.5 percent of the total (Figure 5.4). The distribution of these 660 private patents was:

- Small and medium local enterprises,[4] 80 (or 12.1 percent);
- Large local enterprises, 186 (or 28.2 percent);
- Foreign enterprises, 394 (or 59.7 percent).

---

[2] *Source*: National Survey of R&D in Singapore 2002, Agency for Science, Technology and Research, Singapore.

[3] The term local companies includes wholly 100 percent local-owned companies and companies with more than 30 percent local ownership. Foreign companies include wholly owned foreign companies and those with more than 30 percent foreign ownership. The same rule applies to the innovation survey.

[4] A local small and medium enterprise is defined as one which has total fixed asset not exceeding S$15 million for manufacturing while the additional requirement for non-manufacturing is that the total employment must not exceed 200 workers. A large enterprise is defined as otherwise.

Table 5.1 R&D expenditure in Singapore, 1994–2002

Year	Private sector	Higher education sector	Government sector	Public research institutes	Total R&D expenditure	Total R&D expenditure/ GDP %
	S$ (millions)					
1994	736.2	179.5	142.1	117.2	1,175	1.09
1995	881.4	193.4	110.4	181.4	1,366.6	1.15
1996	1,133.4	238.7	166.8	253.2	1,792.1	1.38
1997	1,314.5	277.7	216.1	296.2	2,104	1.49
1998	1,536.1	305.8	299.8	350.5	2,492.3	1.82
1999	1,670.9	310	304.9	370.6	2,656.3	1.93
2000	1,866	338.3	423.8	381.4	3,009.5	1.91
2001	2,045	367	425.1	395.6	3,232.7	2.13
2002	2,091.3	430	449.1	434.3	3,404.7	2.19

Source: National Survey of R&D in Singapore 2002, Agency for Science, Technology and Research, Singapore.

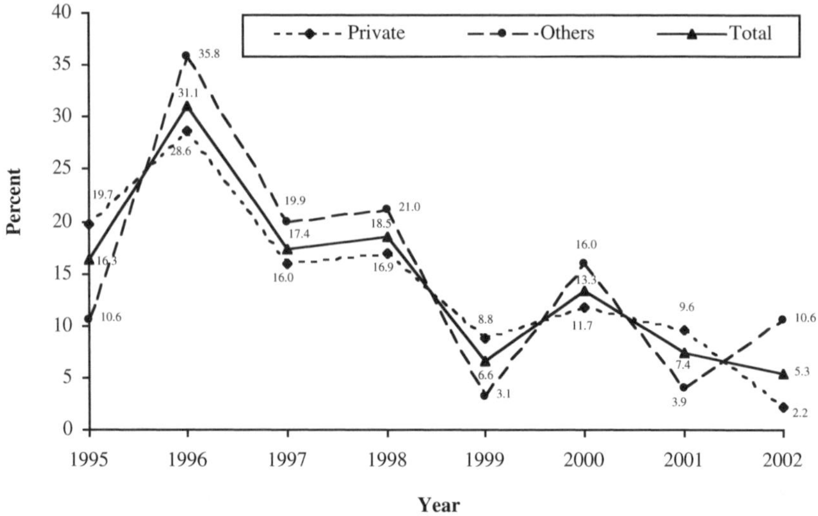

Note: The base year was 1994, therefore the annual percent change in R&D expenditure was from 1995 onwards. The others category includes the government, public research institutes and the higher education sector.

Figure 5.2 Growth rates of R&D expenditure in Singapore, 1994–2002

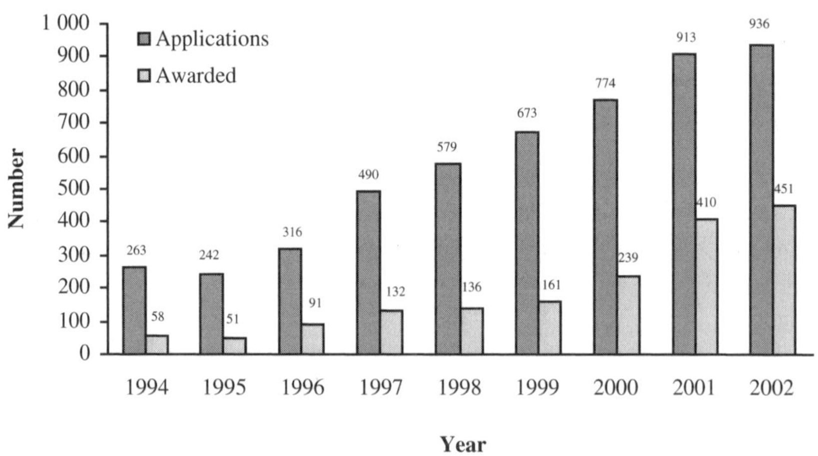

Source: National Survey of R&D in Singapore 2002, Agency for Science, Technology and Research Singapore.

Figure 5.3 Patent applications and awards in Singapore, 1994–2002

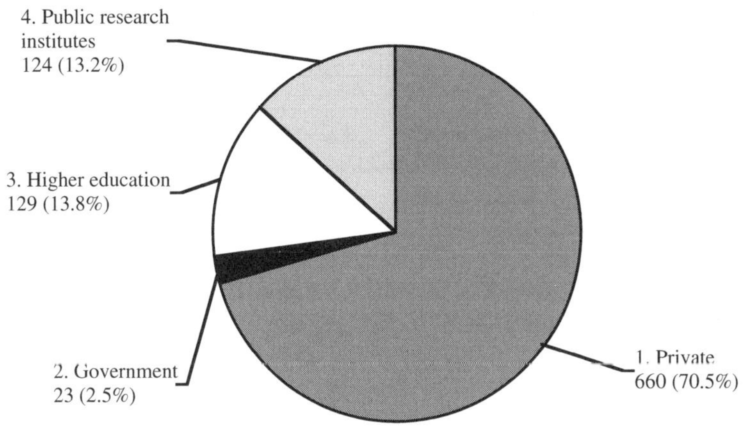

Source: National Survey of R&D in Singapore 2002, Agency for Science, Technology and Research, Singapore.

Figure 5.4 Patent applications in Singapore by source, 2002

This was followed by the public research institutes and higher education sector with 129 (13.8 percent) and 124 patents (13.2 percent) respectively.

Among the local small and medium enterprises, half (41 or 51.3 percent) of the 80 applications belonged to the non-manufacturing sector, which included the R&D, and IT and financial service industries[5] (Table 5.2). The major sources were the natural and life sciences, and IT and IT-related activities. The remaining 39 applications from the manufacturing sector were dominated by firms specialising in machinery and systems and infocomm end products.

By contrast, among the local large enterprises, almost all (183 or 98.3 percent) of the 186 applications were for manufacturing purposes, with the lion's share (152 or 83.1 percent) from the semiconductors industry. The manufacturing sector was just as important to the foreign firms as eight in ten (88 percent or 347) of the 394 patents applied for were for computer peripherals, semiconductors and infocomm end products. The remaining 47 were mostly for IT-related and other service activities from the services sector.

Of 451 patents awarded in 2002, the private sector provided over three-quarters (350 or 77.6 percent) of the total (Figure 5.5). The breakdown of these 350 awards was:

[5] Detailed breakdown in terms of patent applications and awards are from the private sector.

Table 5.2 Private sector patent applications in Singapore by industry, 2002

Applications Singapore, 2002

Description	Local SMEs	Local LEs	Foreign	Total
A – Primary				
Mining and quarrying	0	1	0	1
Sub-total A	**0**	**1**	**0**	**1**
B – Manufacturing				
Biomedical science	1	0	0	1
Semiconductors	1	*152*	*109*	*262*
Computer peripherals	1	0	*114*	*115*
Data storage	0	0	10	10
Infocomm end products	*9*	5	*81*	*95*
Other electronics	0	0	2	2
Speciality chemicals	1	0	5	6
Machinery and systems	*15*	7	9	31
Precision modules	3	0	14	17
Marine and offshore eng	0	6	0	6
Aerospace	0	6	2	8
Land	0	7	1	8
Miscellaneous	8	0	0	8
Sub-total B	**39**	**183**	**347**	**569**
E – Services				
C – Research & development				
Chemicals	0	0	0	0
Electronics	0	0	0	0
Engineering	0	0	1	1
IT	3	0	3	6
Life sciences	2	0	3	5
Other natural sciences	7	0	0	7
Sub-total C	**12**	**0**	**7**	**19**
D – Others				
IT-related	9	2	18	29
Logistics	0	0	0	0
Financial intermediation	9	0	1	10
Other services	11	0	21	32
Sub-total D	**29**	**2**	**40**	**71**
Services total E (C+D)	**41**	**2**	**47**	**90**
Grand total (A+B+E)	**80**	**186**	**394**	**660**

Note: The top-three figures are in italics.

Source: National Survey of R&D in Singapore 2002, Agency for Science, Technology and Research, Singapore.

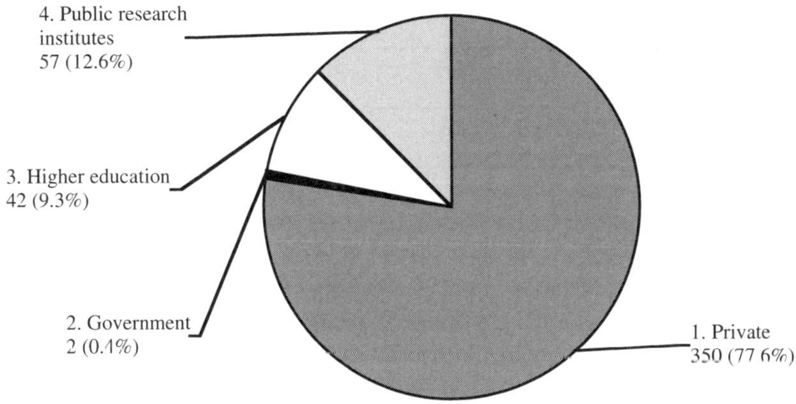

Source: National Survey of R&D in Singapore 2002, Agency for Science, Technology and Research, Singapore.

Figure 5.5 Patent awards in Singapore, 2002

- small and medium local enterprises, 34 (or 9.7 percent)
- large local enterprises, 178 (or 50.8 percent)
- foreign enterprises, 138 (or 39.4 percent).

Next were the public research institutes and the higher education sector with 57 (12.6 percent) and 42 patents (9.3 percent) respectively.

Over half (20 out of 34 or 58.8 percent) of the awards granted to small and medium enterprises belonged to the services sector (Table 5.3). The recipients were mainly in three industries, namely machinery systems, IT and IT-related, and other services. Virtually all (177 of 178 awards or 99.4 percent) of the awards to local large enterprises were for the manufacturing sector, with the semiconductors industry accounting for nine in every ten (161 or 91 percent). The manufacturing sector was also a principal recipient of patent awards for the foreign enterprises (claiming 129 of the 138 awards or 93.5 percent). The awards were less concentrated compared to the local large enterprises. They covered a more diversified range of activities spanning across the semiconductors, computer peripherals, infocomm end products and precision modules industries.

Table 5.3 Private sector patent awards in Singapore by industry, 2002

Description	Local SMEs	Local LEs	Foreign	Total
A – Primary				
Construction	0	1	0	1
Sub-total A	**0**	**1**	**0**	**1**
B – Manufacturing				
Biomedical science	0	0	0	0
Semiconductors	0	*161*	*73*	*234*
Computer peripherals	1	0	*25*	*26*
Data storage	0	0	0	0
Infocomm end products	1	2	*16*	*19*
Other electronics	0	1	2	3
Speciality chemicals	1	0	0	1
Machinery and systems	7	2	1	10
Precision modules	1	0	11	12
Marine and offshore eng	0	1	0	1
Aerospace	0	0	1	1
Land	0	*5*	0	5
Miscellaneous	3	*5*	0	8
Sub-total B	**14**	**177**	**129**	**320**
E – Services				
C – Research and development				
Chemicals	0	0	0	0
Electronics	0	0	0	0
Engineering	1	0	0	1
IT	4	0	0	4
Life sciences	0	0	0	0
Other natural sciences	0	0	0	0
Sub-total C	**5**	**0**	**0**	**5**
D – Others				
IT-related	6	0	4	10
Logistics	0	0	0	0
Financial intermediation	3	0	0	3
Other services	6	0	5	11
Sub-total D	**15**	**0**	**9**	**24**
Services total E (C+D)	**20**	**0**	**9**	**29**
Grand total (A+B+E)	**34**	**178**	**138**	**350**

Note: The top three figures are in italics.

Source: National Survey of R&D in Singapore 2002, Agency for Science, Technology and Research, Singapore.

5.2.2 Innovation

A national survey was conducted by EDB and CMIT[6] in the year 2000 on the level of innovation in Singapore for the year 1999. A total of 371 valid responses (Table 5.4) was obtained from the electronics, chemicals, engineering and life sciences clusters. The objectives of the survey were to:

- assess the level of innovation in the manufacturing sector of Singapore
- benchmark Singapore against OECD countries
- identify problems and challenges to innovation.

Table 5.4 Singapore innovation survey respondents by category

Valid responses	Electronics	Chemicals	Precision & process engineering	Transport engineering	Total
Number of firms	40	69	202	60	371
Percentage	10.8%	18.6%	54.4%	16.2%	100%

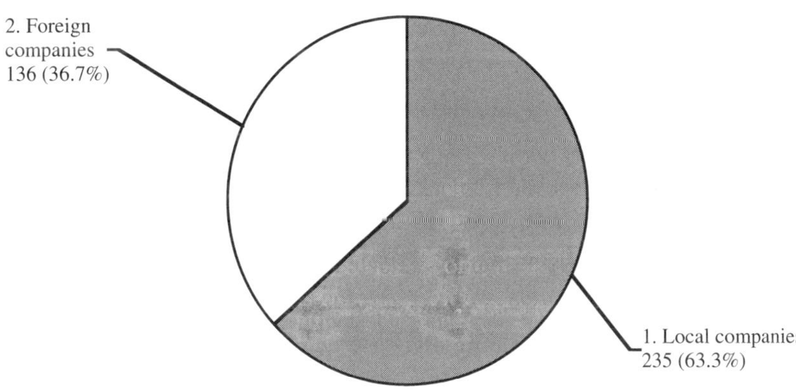

Source: Pattern of Innovation among Singaporean Manufacturing Firms, Centre of Management of Innovation and Technopreneurship (CMIT), National University of Singapore (NUS).

Figure 5.6 Singapore innovation survey respondents by ownership, 2002

[6] Centre of Management of Innovation and Technopreneurship, National University of Singapore. Currently, it is known as NEC or National University of Singapore Entrepreneurship Centre.

OECD's[7] Oslo manual regards a company 'being innovat*ing*' if it has introduced to the market at least one of the following during the past three years:

- a product new to the business or a substantially improved product (product innovation)
- a new or substantially improved production process through new equipment or re-engineering (process engineering).

In addition to this definition, CMIT provides additional conditions for the company where either one of the following criteria below is satisfied to qualify as innovative:

- 25 percent of sales accounted for by new/improved products introduced over the past three years
- 25 percent of production volume accounted for by new/improved processes introduced over the last three years.

The study found the level of innovating[8] firms in Singapore to be 31.6 percent in 1999, while the ratio of innovative firms was smaller at 15.2 percent. The electronics cluster had the highest level of innovation at 56.3 percent, and 54.8 percent of the firms were involved in product and process innovation respectively (Figure 5.7). The transport engineering sector was the lowest in innovation with 14.5 percent innovating firms in product and 12.7 percent in process innovation.

Within the innovating firms, 61.4 percent had up to 24 percent sales derived in 1999 from new/improved products introduced during the past three years between 1996 and 1999 while 19.4 percent of them had sales of 50 percent or more (Table 5.5). Almost half (46.4 percent) of the firms obtained up to 24 percent annual production from new/improved process methods introduced during the past three years. Only about a quarter (23.8 percent) of the firms had production volume of 75 percent or more from innovating processes.

An alternative way of looking at innovation is to observe the degree of product autonomy within the manufacturing sector. It can be proxied by the proportion of sales of the following groups in product design and development:

- manufacturing arm of parent company (MAPC)
- original equipment manufacturing (OEM)

---

7 OECD, 1997: *Oslo Manual, Proposed Guidelines for Collecting and Interpreting Technological Innovation Data*, 2nd edn, Paris.

8 Innovating firms need only to satisfy one of the conditions listed by OECD while the innovative term deemed by CMIT requires the commercial success with at least 25 percent of annual sales or production volume generated by products/processes.

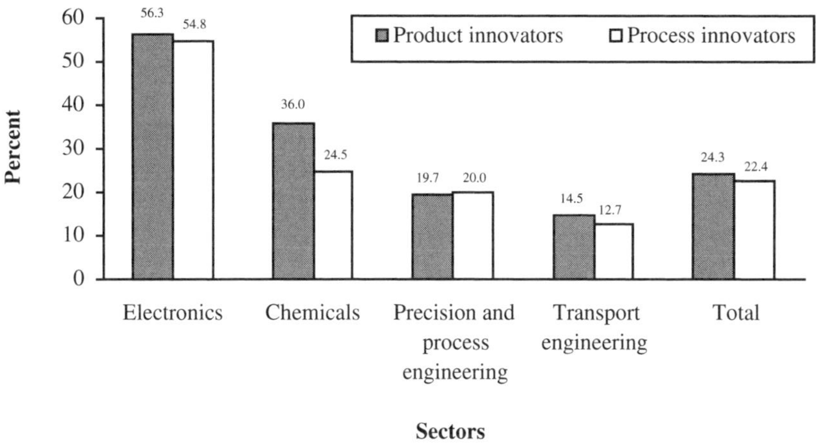

Source: Pattern of Innovation among Singaporean Manufacturing Firms, Centre of Management of Innovation and Technopreneurship (CMIT), National University of Singapore (NUS).

Figure 5.7 Product and process innovating firms in Singapore, 1999

Table 5.5 Singapore innovators by annual sales/production volume, 1999

Description	Electronics	Chemicals	Precision & process engineering	Transport engineering	Total
% of annual sales (1999)					
< 10%	12.5%	35.3%	36.2%	50%	33%
10%–24%	18.8%	41.2%	27.7%	25%	28.4%
25%–49%	31.3%	11.8%	19.1%	12.5%	19.3%
50%–74%	12.5%	11.8%	4.3%	12.5%	8%
≥75%	25%	0%	12.8%	0%	11.4%
Total	**100%**	**100%**	**100%**	**100%**	**100%**
% of Production Volume (1999)					
< 10%	0%	16.7%	24.5%	28.6%	19%
10%–24%	31.3%	41.7%	24.5%	14.3%	27.4%
25%–49%	25%	8.3%	18.4%	14.3%	17.9%
50%–74%	18.8%	25%	8.2%	0%	11.9%
≥75%	25%	8.3%	24.5%	42.9%	23.8%
Total	**100%**	**100%**	**100%**	**100%**	**100%**

- original brand design (OBM)
- original design manufacturing (ODM).

Manufacturing arm of parent company and original equipment manufacturing accounted for $0.5 out of every dollar (49.7 percent) of total sales in 1999. This suggests that manufacturing in Singapore may still be in a stage of value addition rather than value creation.

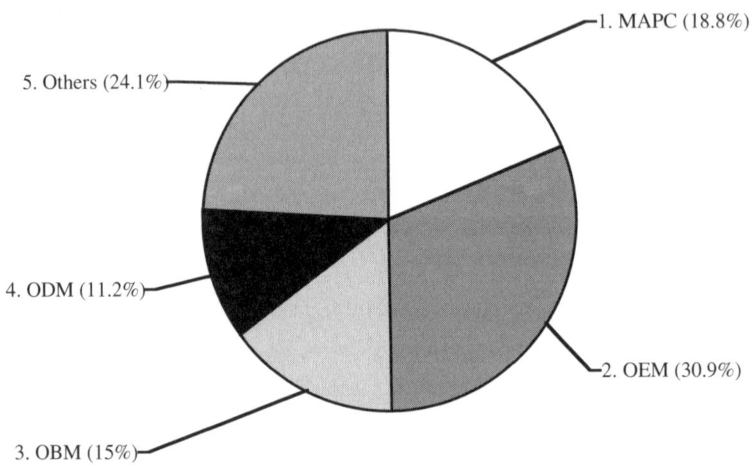

Source: Pattern of Innovation among Singaporean Manufacturing Firms, Centre of Management of Innovation and Technopreneurship (CMIT), National University of Singapore (NUS).

Figure 5.8 Sales by degree of product autonomy in Singapore, 1999

Of the six types of innovative activities, acquisition of machinery (28 percent) was the most popular innovative activity engaged by firms in 1999. Next was R&D (23.2 percent) followed by training linked to technology (21.4 percent). Within the chemicals cluster, R&D was the most cited activity (30.6 percent) while acquisition of machinery (32.3 percent) was more common for the engineering sectors (precision and process and transport).

There were important issues regarding the undertaking or avoidance of innovative new projects by the manufacturing firms. The four most important objectives of respondents in undertaking new innovative projects were:

- opening of new markets and increasing market share
- improving product quality
- reducing production cost
- extending product range.

Table 5.6 Types of innovative activities in Singapore, 1999

Innovative activity	Electronics	Chemicals	Precision & process engineering	Transport engineering	Total
1. Research and development	22.7%	30.6%	22%	19.4%	23.2%
2. Acquisition of R&D services	8%	6.1%	5.1%	9.7%	6.3%
3. Acquisition of machinery	22.7%	24.5%	30.5%	32.3%	28%
4. Licensing of external technology	12%	8.2%	5.6%	9.7%	7.8%
5. Industrial design, market research	12%	10.2%	15.3%	9.7%	13.3%
6. Direct training linked to technology	22.7%	20.4%	21.5%	19.4%	21.4%

Conversely, the four most important reasons for not starting innovative projects were:

- perceived costs too high
- perceived risks too high
- lack of information on technology
- lack of qualified personnel.

Reasons for innovative projects being delayed were:

- lack of qualified personnel
- lack of information on technology
- perceived costs too high
- perceived risks too high.

Within the internal dynamics of the firm (Table 5.7), the three most crucial factors for innovation were:

- top management is highly supportive
- management tolerates failure and encourages learning from mistakes.

Table 5.7 Internal environment for innovation in Singapore

Comment	Mean scores
1. Top management is highly supportive of innovation	2.2
2. Management tolerates failure and encourages staff to learn from mistakes	2.4
3. Employees are very open to changes and new ideas	2.6
4. Management strongly advocates the use of IT in innovating our business processes	2.6
5. We encourage entrepreneurship among our employees	2.8
6. Compensation system is oriented towards rewarding innovative employees	2.9
7. Middle management constantly delivers innovative proposals to the top	2.9
8. We have in place good process for managing innovation projects	3
9. Corporate performance measurement system closely monitors our innovation performance	3

1 = strong agreement, 5 = strong disagreement.

Source: Pattern of Innovation among Singaporean Manufacturing Firms, Centre of Management of Innovation and Technopreneurship (CMIT), National University of Singapore (NUS).

- employees are open to change and new ideas
- management strongly advocates use of IT in innovating business processes.

Table 5.8 depicts the external factors leading to innovation. The four most important factors were:

- telecoms and IT services
- intellectual property protection
- openness of government department and regulatory authorities
- government incentives.

5.2.3 International Comparison

A brief comparison for Singapore and its European counterparts was based on the following:

Table 5.8 External environment for innovation in Singapore

Comment	Mean scores
1. Telecoms and IT services	3.85
2. Intellectual property protection	3.33
3. Openness of government department and regulatory authorities	3.29
4. Government incentives	3.27
5. Openness of customers	3.21
6. Listing requirements of stock exchange	3.2
7. Openness of suppliers	3.14
8. Attitude of people	3.1
9. Consultancy support services	3.08
10. Local universities	3.06
11. Manpower in business sector	3.03
12. R&D institutions	3
13. Other technical supporting services	2.97
14. Venture capital	2.95
15. Scientific–technical manpower	2.89
16. Tolerance for failure	2.64

1 = poor, 5 = good.

Source: Pattern of Innovation among Singaporean Manufacturing Firms, Centre of Management of Innovation and Technopreneurship (CMIT), National University of Singapore (NUS).

- percentage of innovating firms in manufacturing sector
- percentage of product and process innovators in innovating firms
- average proportion of turnover due to new/improved products for product innovators in manufacturing sector.

The proportion of innovating firms at 32 percent in 1999 for Singapore was generally lower than that of the more advanced European nations and it was roughly the same as that of both Belgium and Spain in 1996 (27 percent and 29 percent respectively) (Table 5.9).

The composition of product (24 percent) and process innovators (22 percent) in Singapore's manufacturing sector in 1999 was roughly equivalent to that of both Belgium and Spain in 1996.

The average turnover in 1999 from new/improved products by the product innovators for Singapore was 29 percent. This was higher than France's and

Table 5.9 Innovating manufacturing firms, Singapore vs Europe

Nation	1990–1992	1996
1. Austria	N/A	67%
2. Belgium	61%	27%
3. Denmark	56%	N/A
4. Finland	N/A	36%
5. France	39%	43%
6. Germany[a]	67%	69%
7. Ireland	72%	73%
8. Italy	34%	N/A
9. Luxembourg	37%	42%
10. Netherlands	57%	62%
11. Norway[b]	53%	48%
12. Singapore (1999)	N/A	32%
13. Spain	37%	29%
14. Sweden	N/A	54%
15. United Kingdom	N/A	59%

Notes:
[a] The data for Germany includes Ex-GDR from 1991 for all international comparison from tables IID.7–IID.9.
[b] Norway's figure of 48 percent is for the year 1997.

Source: Pattern of Innovation among Singaporean Manufacturing Firms, Centre of Management of Innovation and Technopreneurship (CMIT), National University of Singapore (NUS).

Netherlands's 21 percent and 25 percent respectively in 1996. However, it was lower than Germany's 43 percent and Ireland's 32 percent.

5.2.4 Benchmarking

An alternative way of benchmarking Singapore's innovativeness on the international arena would be to examine the patent statistics compiled by the OECD based on the following issues (see Figure 5.9):

- number of triadic family patents per billion GDP and million population (relative measure of high-value-added innovation)
- ICT[9]and biotechnology patents as a percentage of individual national total (trend)

[9] ICT stands for information, communication and telecommunications.

Table 5.10 Product and process innovators, Singapore vs Europe

Nation	Product innovators	Process innovators
1. Austria	60%	49%
2. Belgium	24%	17%
3. Finland	29%	25%
4. France	38%	31%
5. Germany	65%	53%
6. Ireland	66%	54%
7. Luxembourg	32%	29%
8. Netherlands	56%	46%
9. Norway (1997)	35%	40%
10. Singapore (1999)	24%	22%
11. Spain	24%	25%
12. Sweden	48%	38%
13. United Kingdom	52%	37%

Note: The figures for the European countries are for 1996 while data for Singapore pertains to the year 1999.

Source: Pattern of Innovation among Singaporean Manufacturing Firms, Centre of Management of Innovation and Technopreneurship (CMIT), National University of Singapore (NUS).

- proportion of foreign and domestic ownership of patents (ownership).

(a) Triadic patent families

The triadic family concept aims to provide for an ever-increasing need for reliable patent statistics and high quality[10] indicators in measuring the performance of the various nations on a comparable basis. The triadic family patent measurement is proposed on the basis that counting patents filed at different offices eliminates the problem of home bias[11] and of ensuring that only high-value-added innovation should be the basis for comparison. A triadic patent family is defined as a set of patents taken in various countries that protects the same innovation that is filed at different patent offices.[12]

---

10 This is based on the assumption that an inventor will file at the different patent offices apart from his/her domestic country after a thorough cost-benefit analysis.

11 The counting of single patent often shows a higher proportion from the domestic side.

12 Currently, OECD's definition of the triad refers to the USPTO, EPO and JPO respectively. Refer to: Dernis, H.D. Guellec and B. van Pottelsberghe (2001), *Using Patents*

Table 5.11 Turnover from new/improved products for product innovators

Nation	Product innovators
1. Austria	31%
2. Belgium	14%
3. Finland	25%
4. France	21%
5. Germany	43%
6. Ireland	32%
7. Netherlands	25%
8. Norway (1997)	20%
9. Singapore (1999)	29%
10. Spain	27%
11. Sweden	31%
12. United Kingdom	23%

Note: Singapore's value was derived from mid-point estimate of the categorised variable. Figures for the European countries are for the year 1996 while data for Singapore pertains to the year 1999.

Source: Pattern of Innovation among Singaporean Manufacturing Firms, Centre of Management of Innovation and Technopreneurship (CMIT), National University of Singapore (NUS).

Singapore's triadic patents per billion \$GDP[13] improved from 0.5 in 1991 to 1.1 in 2000 which was just behind that of the EU (1.6) and the United States (1.7) but far behind Japan's (3.9) (Figure 5.10). Over the same period, Singapore's relative share of such patents in the world increased from 0.07 percent to 0.2 percent.[14]

Singapore's population was more innovative in 2000, with 23.2 triadic patent families per million people, as compared to 6.5 per million people in 1991. Japan was the highest with 92.6 triadic patent families per million population followed by the United States and the EU with 53.1 and 36.2 triadic patents per million population respectively.

Counts for Cross-country Comparisons of Technology Output, STI review no. 27, Paris: OECD.

[13] Gross Domestic Product is based on 1995 US dollars in billions using purchasing power parity.

[14] Singapore is ranked 22 in the year 2000 with a 0.21 percent share which is the same as China. Although Japan has the highest ratio, it is ranked third, based on its share of absolute triadic patents in the world.

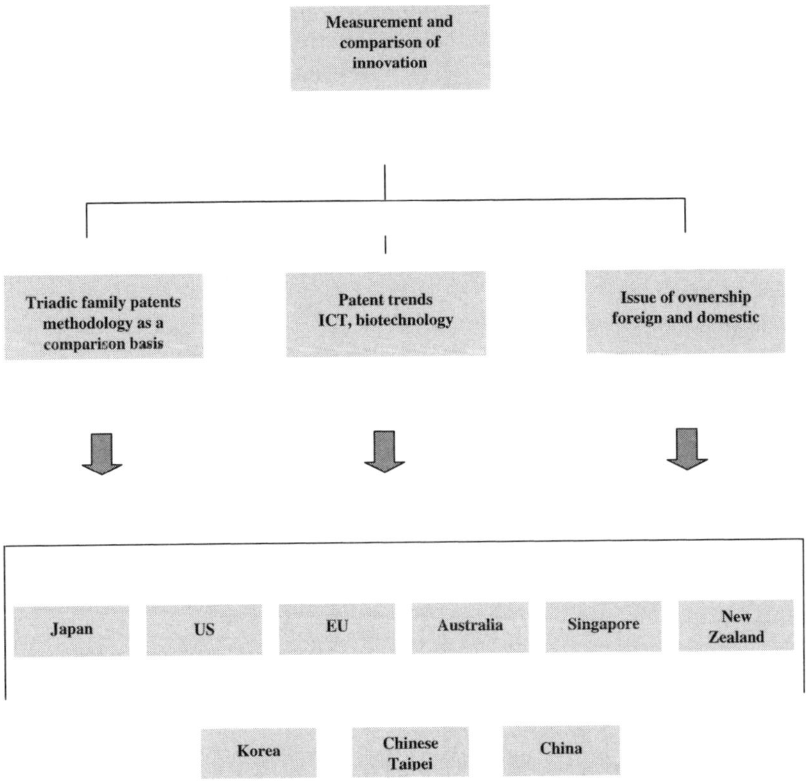

Figure 5.9 Comparison of innovation of selected nations, 1991 and 2000

(b) ICT and biotechnology patents

The second set of statistics involves ICT and biotechnology patents. Their importance was evaluated as proportions of national total patents. The following ICT industries are defined under the ICT[15] patents category:

- telecommunications
- consumer electronics
- computers, office machinery
- other ICT.

[15] IPC is the international patent classification. Refer to www.wipo.int/classifications/fulltext/new_ipc/index.html.

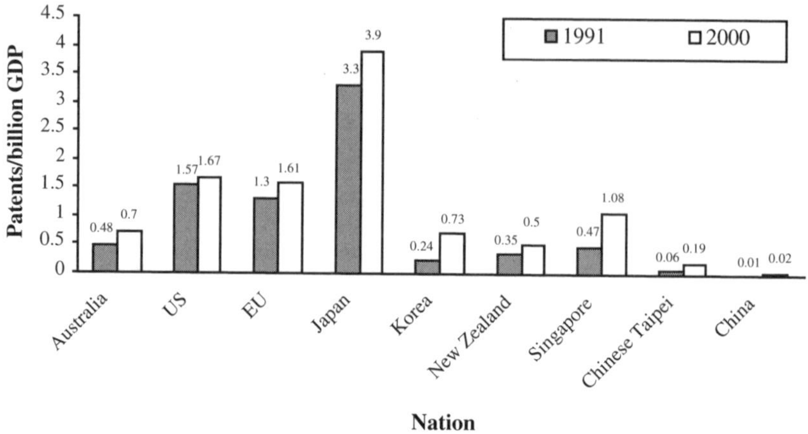

Source: Compendium of Patent Statistics, 2004, OECD.

Figure 5.10 Triadic family patents per billion GDP, 1991 and 2000

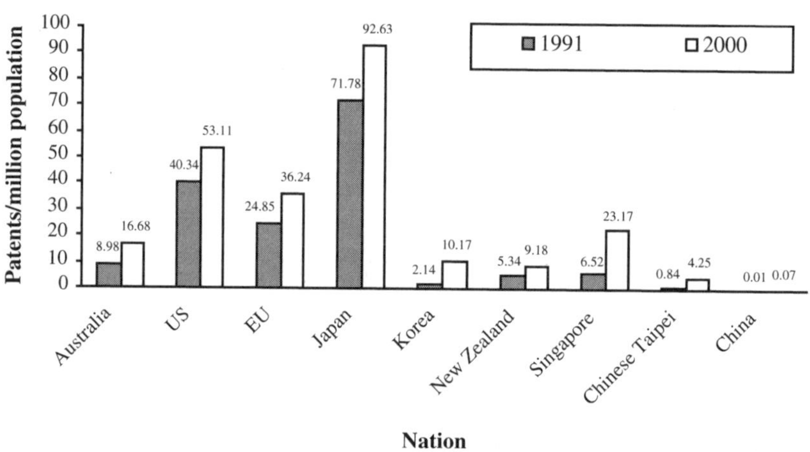

Source: Compendium of Patent Statistics, 2004, OECD.

Figure 5.11 Triadic family patents per million population, 1991 and 2000

ICT patents in Singapore formed an integral part of its own national total as they rose from 48.4 percent to 62.6 percent between 1991 and 2000. Singapore's share in 2000 was the highest in the world. All except Japan experienced increasing shares of ICT patents between 1991 and 2000. Japan recorded a slight fall of 1.8 percent points from 45.9 percent in 1991 to 44.1 percent in

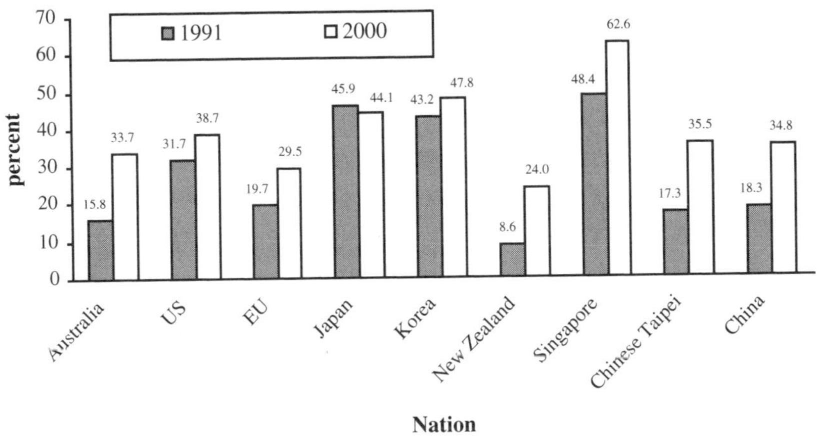

Source: Compendium of Patent Statistics, 2004, OECD.

Figure 5.12 ICT patents as a percentage of national total, 1991 and 2000

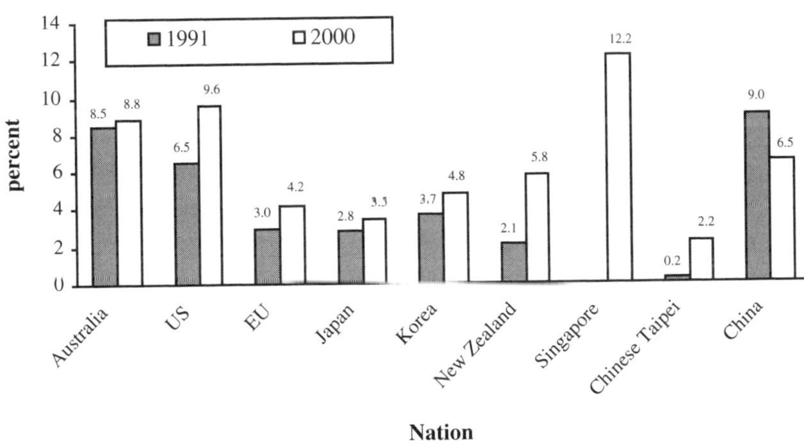

Source: Compendium of Patent Statistics, 2004, OECD.

Figure 5.13 Biotechnology patents as a percentage of national total, 1991 and 2000

2000. By contrast China, Taiwan (Chinese Taipei) and New Zealand each witnessed almost a two-to-three-fold increase in their ICT share in total patents.

In the case of biotechnology patents, Singapore's share in 2000 at 12.2 percent was also the highest in the world[16] (Figure 13). Except for China, all the other seven selected countries experienced an increase in their proportions of biotechnology patents in respective national total of patents between 1991 and 2000. China's share fell from 9 percent in 1991 to 6.5 percent in 2000.

(c) Foreign ownership of patents

The third benchmarking indicator is ownership of patents. The high level of foreign ownership of patents in Singapore reflects its open economy. The share of foreign patents has declined from 79.4 percent in 1991 to 50.3 percent in 2000 which was still higher than the other nations except China (Figure 5.14). In Japan and Korea, foreign ownership amounted to less than 10 percent in 2000. The level of domestic ownership of inventions abroad for Singapore improved from 21.7 percent in 1991 to 28.6 percent in 2000 (Figure 5.15). Among the nine selected countries, China had the greatest increase of 9.9 percentage points from 12.8 percent to 22.7 percent. Only New Zealand and Korea witnessed a decrease in domestic patents abroad in 2000, as against that in 1991.

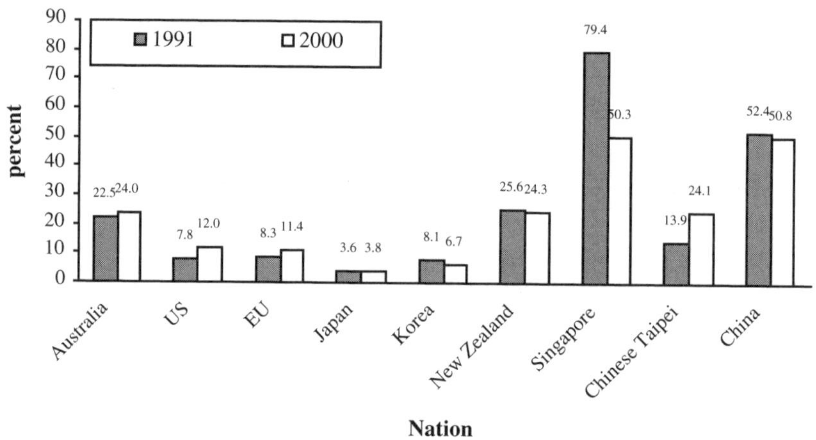

Source: Compendium of Patent Statistics, 2004, OECD.

Figure 5.14 Foreign ownership in domestic inventions, 1991 and 2000

The level of domestic ownership of inventions abroad for Singapore improved from 21.7 percent to 28.6 percent between 1991 and 2000. Among them, China

16 Comparative data for Singapore is not available for the year 1991.

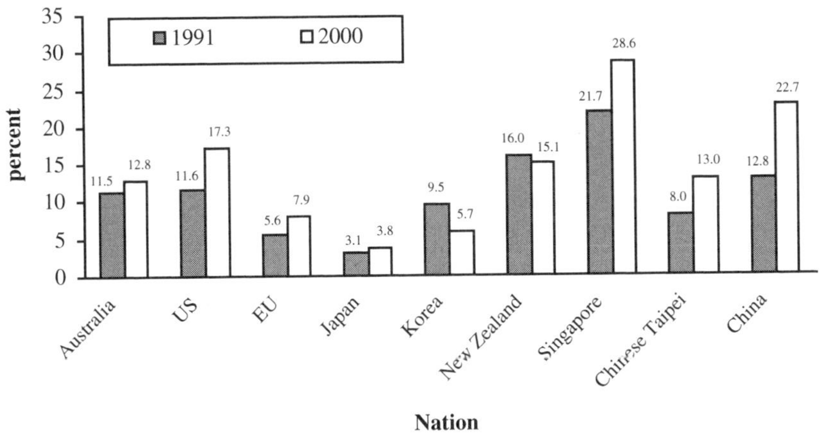

Source: Compendium of Patent Statistics, 2004, OECD.

Figure 5.15 Domestic ownership in inventions made abroad, 1991 and 2000

had the greatest increase of 9.9 percent from 12.8 percent to 22.7 percent. Only New Zealand and Korea witnessed a slight decrease from 16 percent to 15.1 percent during this time.

5.3 SURVEY FINDINGS IN SINGAPORE

The survey information from the sample of 104 firms (47 in products or manufacturing and 57 in services) cover:

- demographic profile e.g. ownership, employment size
- level of innovation
- usage of the intellectual property system in Singapore and abroad
- views on the necessity of protecting minor inventions
- UMs as an incentive in encouraging innovations.

5.3.1 Demographic Profile

A majority of the respondents are local firms i.e. local equity of more than 30 percent.[17] Two out of three firms (67 percent) are locally owned. Manufacturing

[17] We follow the A*Star classification where local firms include wholly 100 percent local and those with more than 30 percent local ownership.

firms tend to have proportionately more wholly foreign-owned firms (40 percent) than services firms (23 percent).

Four-fifths (80.2 percent) of the firms employed up to 250 employees at the time of the survey, indicating that the majority of the firms are within the small to medium sized category. About one in eight firms (12 percent) employed between 251 and 1000 workers while the remaining one-thirteenth (8 percent) had over 1000 workers. As expected, firms in the services sector tend to be small. Over half (54 percent) of service firms engaged 25 or fewer workers, whereas two-fifths (42 percent) of manufacturing firms are of this size.

Half (49 percent) of the firms had sales below S$5 million in 2003. In particular, one in five firms (20 percent) is very small with annual sales of less than S$0.1 million. At the other extreme, very large firms with annual sales of at least S$100 million constituted one-quarter (24 percent) of the firms surveyed.

Corresponding to employment, the services sector tends to have proportionately more small firms than the manufacturing sector. Half (53 percent) of the

Table 5.12 Profile of respondents

Profile and demographics	Products	%	Services	%	Total	%
(i) Ownership						
Wholly local	22	46.8	34	59.6	56	53.8
≥ 30 local	5	10.6	9	15.8	14	13.5
≤ 30% local	1	2.1	1	1.8	2	1.9
Wholly foreign	19	40.4	13	22.8	32	30.8
Total	**47**	**100**	**57**	**100**	**104**	**100**
(ii) No. of full-time employees						
0–25	19	42.2	30	53.6	49	48.5
26–100	9	20.0	10	17.9	19	18.8
101–250	8	17.8	5	8.9	13	12.9
251–500	4	8.9	5	8.9	9	8.9
501–1000	2	4.4	1	1.8	3	3.0
1001 and above	3	6.7	5	8.9	8	7.9
Total	**45**	**100**	**56**	**100**	**101**	**100**
(iii) Annual sales in 2003						
<S$ 0.1 million	9	20.5	10	18.9	19	19.6
S$0.1 million–S$0.9 million	4	9.1	8	15.1	12	12.4
S$1 million-S$4.9 million	6	13.6	10	18.9	16	16.5
S$5 million–S$9.9 million	2	4.5	5	9.4	7	7.2
S$10 million–S$49.9 million	8	18.2	8	15.1	16	16.5
S$50 miillion–S$99.9 million	1	2.3	3	5.7	4	4.1
S$100 miillion–S$499.9 million	8	18.2	5	9.4	13	13.4
S$500 million and above	6	13.6	4	7.5	10	10.3
Total	**44**	**100**	**53**	**100**	**97**	**100**

firms in the services sector obtained sales of less than S$5 million in 2003, compared to two-fifths (43 percent) among manufacturing firms. Very large firms with sales of at least S$100 million accounted for one in six (17 percent) service firms and one in three (32 percent) manufacturing firms.

Half (55 percent) of the firms have expanded abroad. Malaysia is the most popular country, followed by China.

5.3.2 Level of Innovation

One-fifth (18.8 percent) of the firms are highly innovative and proportionately more of them are in the manufacturing sector. One in two firms (47.5 percent) is moderately innovative while one in five (19.8 percent) is at the start-up level and one in seven (13.9 percent) is not innovative. On the whole, two-thirds of the firms are at least moderately innovative.

Three-quarters of the firms have designed at least one innovative product/service in the past three years. Almost half (46 percent) of the firms have between one and six innovations over the same period, while three-tenths (29 percent) have more than six innovations. Of the remaining one-fourth (25 percent) with no innovations within the past three years, proportionately more are from the services sector.

Half of the firms have up to 24 percent of their revenue in 2003 generated by new or improved products/services introduced during the past three years. One in five (22 percent) has from 24 to less than 75 percent of its revenue derived from innovations. And one-tenth of the firms has innovations contributing to at least 75 percent of revenue. By contrast, a small proportion (15 percent) of the respondents has no revenue generated from innovations. Again proportionately more firms in the manufacturing sector have larger shares of revenue from innovations than in the services sector. Two-fifths (41 percent) of manufacturing firms received 25 percent or more of their revenue from innovative products/services in the past three years, while similar firms constituted three-tenths (30 percent) in the services sector.

5.3.3 Data on Usage of Intellectual Property System

About half (45 percent) of the firms used the intellectual property system in Singapore. Interestingly proportionately more firms in the services sector make use of Singapore's IP system than manufacturing firms (49 percent vs 41 percent for manufacturing).

To the multiple-response question on six specified types of IP used, 46 respondents provided a total of 93 mentions, giving a ratio of two types of IP usage per firm that engages in IP activity. The most popular IP used, as measured by the proportion of firms using the IP out of the total number of firms which

Table 5.13 Level of innovation

Innovation	Products	%	Services	%	Total	%
(i) Level of innovation						
Start-up	9	19.1	11	20.4	20	19.8
Not innovative	5	10.6	9	16.7	14	13.9
Moderately innovative	22	46.8	26	48.1	48	47.5
Highly innovative	11	23.4	8	14.8	19	18.8
Total	**47**	**100**	**54**	**100**	**101**	**100**
(ii) No. of innovative products/services in past 3 years						
0	10	21.3	15	28.8	25	25.3
1–6	26	55.3	19	36.5	45	45.5
>6	11	23.4	18	34.6	29	29.3
Total	**47**	**100**	**52**	**100**	**99**	**100**
(ii) % of revenue consisting of new products/services or improvement on existing sales introduced by company over the past 3 years						
0%	6	13.6	8	16.0	14	14.9
<10%	8	18.2	15	30.0	23	24.5
10%–24%	12	27.3	12	24.0	24	25.5
25%–49%	7	15.9	7	14.0	14	14.9
50%–74%	4	9.1	3	6.0	7	7.4
≥75%	7	15.9	5	10.0	12	12.8
Total	**44**	**100**	**50**	**100**	**94**	**100**

responded to the question, are registered trademarks and patents. Six in ten of the respondents have registered trademarks (61 percent) and patents (59 percent). This is followed by copyrights (39 percent), trade secrets (20 percent), registered designs (15 percent) and layout designs (9 percent).

A total of 94 answers was obtained from 56 firms which reported their reasons for not using the IP system. The three most common reasons, as cited by over one-fifth of the firms, are cost of registration/litigation (28.6 percent), intellectual property is not important (26.8 percent), and complexity of legal rules and decision (25 percent). These are followed by complexity of procedure/application (18 percent) and protection too narrow against imitation (16 percent). Less than one in ten respondents avoided using the IP system because of inconvenience (9 percent), short service life-span (7 percent) and registration delays (4 percent).

Others reasons given include:

Table 5.14 Usage of Singapore intellectual property system

Intellectual property system	Products	%	Services	%	Total	%
(i) Does your company make use of the intellectual property system in Singapore?						
Yes	19	40.4	27	49.1	46	45.1
No	28	59.6	28	50.9	56	54.9
Total	**47**	**100**	**55**	**100**	**102**	**100**
(ii) Type of intellectual property used? (multiple responses)						
Patents	11	57.9	16	59.3	27	58.7
Copyright	8	42.1	10	37.0	18	39.1
Layout design	1	5.3	3	11.1	4	8.7
Registered design	2	10.5	5	18.5	7	15.2
Registered trademark	12	63.2	16	59.3	28	60.9
Trade secrets	5	26.3	4	14.8	9	19.6
(iii) What are the reasons for not using the intellectual property system? (multiple responses)						
Cost of registration/litigation	7	25.0	9	32.1	16	28.6
Complexity of procedure/ application	4	14.3	6	21.4	10	17.9
Protection too narrow against imitation	5	17.9	4	14.3	9	16.1
Service life-span	2	7.1	2	7.1	4	7.1
Delays in registration	1	3.6	1	3.6	2	3.6
Inconvenience	4	14.3	1	3.6	5	8.9
Complexity of legal rules and decisions	7	25.0	7	25.0	14	25.0
Intellectual property protection is not important	7	25.0	8	28.6	15	26.8
Others	11	39.3	8	28.6	19	33.9

- Intellectual property is not relevant or applicable to business
- Intellectual property affairs are carried out at head office overseas
- Sales are outside Singapore
- Lack of knowledge and time.

Coincidentally about half (45 percent) of the respondents have made use of overseas intellectual property systems. This is the same proportion as for the Singapore IP system. But unlike the local system, proportionately more manufacturing firms are users of IP than services firms.

Nine out of 20 respondents (45.3 percent) made use of overseas intellectual property systems while the remaining 11 (54.7 percent) did not. On the use of foreign intellectual property systems, 43 firms responded with a total number of 67 mentions on the types of usage. The proportion of firms using each form of overseas intellectual property system was:

- patents (72.1 percent)
- trademarks (53.5 percent)
- copyright (18.6 percent)
- utility models (9.3 percent)
- others (2.3 percent).

Table 5.15 Usage of overseas intellectual property systems

Overseas intellectual property system	**Products**	**%**	**Services**	**%**	**Total**	**%**
(iv) Does your company currently make use of overseas intellectual property systems?						
Yes	22	50.0	21	41.2	43	45.3
No	22	50.0	30	58.8	52	54.7
Total	**44**	**100**	**51**	**100**	**95**	**100**
(v) Type of overseas intellectual property used? (multiple responses)						
Copyright	6	27.3	2	9.5	8	18.6
Patents	19	86.4	12	57.1	31	72.1
Trademarks	10	45.5	13	61.9	23	53.5
Utility models	2	9.1	2	9.5	4	9.3
Others	0	0.0	1	4.8	1	2.3

The pattern for overseas IP differs from that of local IP. An average of 1.6 types of IP is used by each firm that uses overseas IP systems.[18] Overseas patents are the most popular with seven in ten firms (72 percent) using it, followed by

[18] A total of 67 mentions was made by 43 firms.

trademarks (54 percent). These are larger proportions than that for Singapore patents.

Overseas, copyright (used by 19 percent of the firms) is less popular than at home. Of special interest to this study is that one in ten firms (9 percent) that use overseas IP makes use of UMs abroad. One in three (35 percent) of the 89 respondents feels that legal protection increases the company's revenue. Larger proportions of manufacturing firms (48 percent) experienced higher revenue from IP protection than service firms (23 percent).

Indications of the extent of revenue increase suggest that half (46 percent) of the firms which benefited from IP protection experienced higher revenue of less than 10 percent. The revenue of almost two-fifths of the firms rose by 10 percent to 24 percent.

One in ten firms (8 percent) each experienced revenue escalating by 25 percent–49 percent and 50 percent–74 percent respectively. There are larger proportions of service firms with strong growth in revenue from IP protection than manufacturing firms. Two-thirds (67 percent) of service firms indicated revenue increase of at least 10 percent as against half (47 percent) of manufacturing firms.

Table 5.16 Protection of law and impact on revenue

Protection of law and impact on revenue	**Products**	%	**Services**	%	**Total**	%
(vi) Has the protection of the law increased the revenue of your company?						
Yes	20	47.6	11	23.4	31	34.8
No	22	52.4	36	76.6	58	65.2
Total	**42**	**100**	**47**	**100**	**89**	**100**
(vii) By how much has the protection of the law increased the revenue of your company?						
<10%	8	53.3	3	33.3	11	45.8
10%–24%	5	33.3	4	44.4	9	37.5
25%–49%	2	13.3	0	0.0	2	8.3
50%–74%	0	0.0	2	22.2	2	8.3
Total	**15**	**100**	**9**	**100**	**24**	**100**

One-third (34 percent) of the respondents have encountered copying in the minor products/ services of their company. A larger proportion (39 percent) of service firms is faced with this problem than manufacturing firms (30 percent). On the extent of copying in the respondent's industry, one in five firms (18 percent) feels that there is no copying within their industry. On the other hand, one in five firms (17 percent) views copying in their industry to be massive. The rest are equally split between moderate copying (33 percent) and a small amount of copying (32 percent).

There tends to be more extensive copying in the service industries than in manufacturing industries. Three-fifths (58 percent) of service firms assessed copying in their industries as moderate to massive, while two-fifths (41 percent) of manufacturing firms judged industry copying to be moderate to massive. A large majority (72 percent) of the respondents believe that inventions, no matter how minor, should be protected against imitation. Proportionately more manufacturing companies hold this view (80 percent vs 66 percent among service firms) though copying is more extensive in service industries than in manufacturing industries as just observed.

Table 5.17 Protection of minor innovations

Imitation and protection of minor innovations	Products	%	Services	%	Total	%
(viii) Has any minor services/products of your company been copied?						
Yes	13	29.5	19	38.8	32	34.4
No	31	70.5	30	61.2	61	65.6
Total	**44**	**100**	**49**	**100**	**93**	**100**
(ix) Do you feel there is massive copying in your industry?						
No	10	21.7	8	15.4	18	18.4
Small	17	37	14	26.9	31	31.6
Moderate	14	30.4	18	34.6	32	32.7
Massive	5	10.9	12	23.1	17	17.3
Total	**46**	**100**	**52**	**100**	**98**	**100**
(x) Do you feel that inventions, no matter how minor should be protected against imitation?						
Yes	36	80	35	66	71	72.4
No	9	20	18	34	27	27.6
Total	**45**	**100**	**53**	**100**	**98**	**100**

Four-fifths (83 percent) of the respondents indicated a need for a large and free public domain to assist follow-on innovation. This large majority view is held across manufacturing and service firms.

Table 5.18 Need for public domain

Public domain	**Products**	**%**	**Services**	**%**	**Total**	**%**
(xi) Do you feel there is a need for a large and free public domain to assist follow-on innovation?						
Yes	34	85	40	81.6	74	83.1
No	6	15	9	18.4	15	16.9
Total	**40**	**100**	**49**	**100**	**89**	**100**
(xii) Do you feel there is a need for Singapore to acquire a more rapid and cheaper protection of innovation in order to encourage industrial growth and promotion?						
Yes	37	92.5	42	95.5	79	94
No	3	7.5	2	4.5	5	6
Total	**40**	**100**	**44**	**100**	**84**	**100**

5.3.4 Need for UMs?

Half (53 percent) of the firms responded that they would use the UM system to protect their innovations. Service firms are more likely to use the utility model than manufacturing firms. Proportionately more service firms (57 percent) indicated usage of UMs to protect their innovations than manufacturing firms (48 percent).

Again half (47 percent) of the firms would be induced to develop more innovative products/services if a UM were introduced. The likelihood of more innovative activities is greater among services firms than manufacturing firms. Half (52 percent) of service firms would come up with more innovative services if a UM were adopted, compared with two-fifths (41 percent) among manufacturing firms.

One-fifth (20 percent) of the firms would be induced to increase innovative products/services by 50 percent or more. Proportionately more service firms (25 percent) would be induced to expand innovation by at least 50 percent compared to manufacturing firms (13 percent). At the lower end of innovation

Table 5.19 Utility model and innovation

Utility model and its impact	Products	%	Services	%	Total	%
(i) Would you use your utility model to protect your innovation?						
Yes	20	47.6	30	56.6	50	52.6
No	22	52.4	23	43.4	45	47.4
Total	**42**	**100**	**53**	**100**	**95**	**100**
(ii) Would your company be induced to come up with more innovative products/ services if the utility model is adopted?						
Yes	17	40.5	26	52.0	43	46.7
No	25	59.5	24	48.0	49	53.3
Total	**42**	**100**	**50**	**100**	**92**	**100**
(iii) Would your company be induced to come up with more innovative products/ services if the utility model is adopted? By how many % more?						
<10%	6	37.5	7	29.2	13	32.5
10%–24%	4	25.0	8	33.3	12	30.0
25%–49%	4	25.0	3	12.5	7	17.5
50%–74%	2	12.5	5	20.8	7	17.5
≥75%	0	0.0	1	4.2	1	2.5
Total	**16**	**100**	**24**	**100**	**40**	**100**
(iv) If the utility models were adopted, how many applications would you likely file per year?						
0	14	41.2	12	28.6	26	34.2
1–2	14	41.2	22	52.4	36	47.4
3–4	5	14.7	6	14.3	11	14.5
≥5	1	2.9	2	4.8	3	3.9
Total	**34**	**100**	**42**	**100**	**76**	**100**

expansion, about one-third of firms would increase innovation by less than 10 percent and another third by 10 percent–24 percent.

If a UM regime were adopted, half (47 percent) of the firms would file up to two utility model applications per year while one-third (34 percent) would not file any. Service firms are more likely to apply for utility models than manufacturing firms as reflected in their lower proportion of no-filing responses (29 percent vs 41 percent for manufacturing).

One in two (50 percent) respondents felt that the introduction of a UM system would increase their revenue. The positive outlook is slightly more pervasive among service firms (53 percent) than among manufacturing firms (46 percent). Four out of five (78 percent) respondents felt that utility models would help increase their revenue by up to 24 percent. Consistent with the more positive view among service firms, half predict a revenue increase of 10–24 percent, compared to one-quarter among manufacturing firms.

With neighbours having UM systems, would Singapore be disadvantaged if it continues not to have one? Three-fifths (62 percent) of the firms responded to the question that it would be. The same proportion of service and manufacturing firms holds this view.

Table 5.20 Likely effects of utility models on revenue

Effects of utility model on revenue	Products	%	Services	%	Total	%
(v) Do you think the utility model would increase the revenue of your company?						
Yes	18	46.2	24	53.3	42	50.0
No	21	53.8	21	46.7	42	50.0
Total	**39**	**100**	**45**	**100**	**84**	**100**
(vi) Do you think the utility model would increase the revenue of your company? By how much?						
<10%	8	50.0	7	29.2	15	37.5
10%–24%	4	25.0	12	50.0	16	40.0
25%–49%	2	12.5	5	20.8	7	17.5
50%–74%	2	12.5	0	0.0	2	5.0
≥75%	0	0.0	0	0.0	0	0.0
Total	**16**	**100**	**24**	**100**	**40**	**100**
(vii) Would you feel disadvantaged if Singapore does not have a utility model system and yet her neighbours have one?						
Yes	24	63.2	31	60.8	55	61.8
No	14	36.8	20	39.2	34	38.2
Total	**38**	**1**	**51**	**1**	**89**	**1**

5.3.5 Statistical Testing

Statistical tests are conducted to find out whether there is any significant difference in the following:

1. level of innovation
2. degree of copying in industry
3. usage of local/foreign intellectual property systems
4. adoption of UMs upon introduction
5. creation of new products/services upon introduction of UM system

with respect to:[19]

- activity i.e. between manufacturing and service firms
- ownership i.e. between foreign and local firms
- size i.e. between small and large firms.

The responses to the level of innovation that best described the firm are converted to scores as follows:

1 = at start-up
2 = not innovative
3 = moderately innovative
4 = highly innovative.

The test results show that there is a significant difference in the level of innovation by ownership and size. Foreign firms are more innovative than local firms. Large firms are more innovative than small firms as their respective mean scores of 2.90 and 2.50 are significantly different at 5 percent level.

The degree of copying in the industry is represented by the following scores:

1 = none
2 = a small amount
3 = a moderate amount
4 = a massive amount.

[19] For the purposes of statistical testing, local firms consist of wholly owned local and those more than 30% local ownership while large firms are those with S10 million and above in annual turnover.

Table 5.21 Innovation and copying

Issues	(i) Level of innovation			(ii) Degree of copying		
	Number	Mean	*P*-value	Number	Mean	*P*-value
Activity						
Products	47	2.74		46	2.30	
Services	54	2.57	0.397	52	2.65	0.080**
Total	**101**	**2.65**		**98**	**2.49**	
Ownership						
Local	69	2.54		65	2.54	
Foreign	32	2.91	0.085**	33	2.39	0.496
Total	**101**	**2.65**		**98**	**2.49**	
Size						
Small	52	.50		50	2.46	
Large	42	2.90	0.045*	42	2.50	0.848
Total	**94**	**2.68**		**92**	**2.48**	

* Denotes significant difference at 95 percent confidence level.
** Denotes significant difference at 90 percent confidence level.

Copying is greater in service industries than in manufacturing industries. Service firms have a significantly higher mean score of 2.65 as compared to 2.30 for manufacturing firms at 10 percent significance level.

Concerning IP systems (Table 5.22), the results show that usage of the local IP system is independent of activity, ownership and size of firms. None of the proportions are significantly different from the expected proportions of equal distribution.

The use of foreign IP systems is dependent on ownership and size. The proportion of foreign firms using overseas IP systems is higher than that of local firms at 5 percent significance level. Large firms are more likely to use foreign IP systems as their proportion is significantly higher than small firms at the 5 percent significance level.

The usage of UMs is not affected by type of activity of firm (manufacturing vs services), ownership (foreign vs local) and size (small vs large).

Similarly, the incentive to come up with new products/services upon the UM system being introduced was independent of activity, ownership and size.

Table 5.22 Intellectual property systems

Issues	Usage of local intellectual property system			Usage of foreign intellectual property system		
	Yes	**No**	***P*-value**	**Yes**	**No**	***P*-value**
Activity						
Products	19	28		22	22	
Services	27	28	0.381	21	30	0.389
Total	**46**	**56**		**43**	**52**	
Ownership						
Local	32	37		22	44	
Foreign	14	19	0.707	21	8	0.000*
Total	**46**	**56**		**43**	**52**	
Size						
Small	25	27		16	33	
Large	19	24	0.705	23	17	0.019*
Total	**44**	**51**		**39**	**50**	

* Denotes significant difference at 95 percent confidence level.
** Denotes significant difference at 90 percent confidence level.

5.4 THE CASE STUDIES

5.4.1 Company A – A Healthy Lifestyle Products Company

Company A is a Singapore public company listed on the Singapore Stock Exchange Mainboard and is a global leader in healthy lifestyle products. Company A manufactures and sells a wide range of healthy lifestyle products focusing on areas such as health, hygiene, nutrition and fitness. Company A was established in 1980 and started off as a trading company selling household goods. It later ventured into healthy lifestyle products and started to sell handheld massages and foot reflexology rollers. Now, Company A has more than 500 outlets in Asia, the Middle East, United Kingdom and North America. Its annual turnover for 2003 was S$287.4 million and more than 80 percent of this was generated by markets outside of Singapore. Company A owes much of its success to innovative marketing strategies and a constant drive to produce successful products with superior designs, features and quality.

Table 5.23 Utility models

Issues	Using utility model system if it is introduced			New products/services if utility model is adopted		
	Yes	**No**	***P*-value**	**Yes**	**No**	***P*-value**
Activity						
Products	20	22		17	25	
Services	30	23	0.384	26	24	0.270
Total	**50**	**45**		**43**	**49**	
Ownership						
Local	34	30		31	30	
Foreign	16	15	0.890	12	19	0.271
Total	**50**	**45**		**43**	**49**	
Size						
Small	24	26		24	24	
Large	23	15	0.243	18	20	0.808
Total	**47**	**41**		**42**	**44**	

* Denotes significant difference at 95 percent confidence level.
** Denotes significant difference at 90 percent confidence level.

In 2002, Company A set up an in-house R&D team consisting of product design research and technology research. The R&D team is tasked to explore new technologies, product platforms and hybrid products. Since its inception, the R&D team in Company A has launched several innovative products including massage chairs, blood pressure meters, eye massagers and scales that read and calculate weight, body fat and water content.

Company A is a strong believer in owning and developing intellectual property. It has more than 10 registered design rights under its name for the various innovative designs of its very successful products. It does not own any patents at the moment but is in the process of filing two UMs in more than 10 countries in the world. The filing of these patents overseas is done via the Patent Cooperation Treaty (PCT) route which is more cost-effective and time saving.

Company A's current strategy is to seek and secure protection for its innovative products and designs in Singapore and elsewhere in the world. The main objective is to erect entry barriers to its rivals and deter illegitimate copying of its innovative designs. As Company A's present competitive edge lies mainly in the innovative designs of its healthy lifestyle products, it uses the registered designs protection system in Singapore and elsewhere to maintain this competitive edge. As Company A is gradually building up its technology and functional capabilities, it is now also looking into filing patents to maintain its lead in the relevant art. As a strategy, Company A also actively and vigilantly enforces its intellectual property rights. It has taken legal action in both Taiwan and China for counterfeits and substantial illegitimate copying of its designs in massage chairs.

On the whole, Company A feels that designs and patent registrations help to increase the company's competitive edge. With the legal protection granted by the government, Company A seeks to deter others and this gives the company a head-start in the exploitation and commercialisation of its innovative products. Company A notes that as patent and design registrations often involve the complete disclosure of the invention concerned to the Patent Office and the public, such a protection scheme may not always be the most appropriate form of protection. In fact, from a strategy point of view, it may be more advantageous to resort to secrecy particularly for nutritional and health supplement products.

Besides registered designs and pending patent applications, Company A also owns other intellectual property rights such as trade marks and copyrights.

What are Company A's views on the need for an alternative protection system for minor inventions and innovations such as the UM system in Singapore? In fact, Company A has already begun the filing processes of two UMs outside of Singapore and therefore it is familiar with the UM systems already implemented in many developed countries. Company A thinks that the current standard patent system has its merits but it is less responsive to the needs of smaller inventions with a shorter product lifespan. Such inventions also require protection to deter copying as they are even more vulnerable to illegitimate copying since they usually involve a simple but innovative idea. It will be a boost to many companies' competitive edge if such inventions can also secure quick and cost-effective legal protection from the government. Company A is aware that several of our neighbours in the region have already adopted UM systems and it feels disadvantaged as Singapore does not have such an alternative protection system. Therefore, Company A thinks that there is a need for Singapore to look into the possible implementation of an alternative protection system for minor inventions and innovations.

As just noted, more than 80 percent of Company A's annual turnover is generated by markets outside of Singapore. Consequently, Company A is more interested to secure protection in markets outside of Singapore. It has however, voiced concerns over the expense required to obtain protection at an international level and to enforce its rights in the various jurisdictions. Nevertheless, Company A is committed to its strategy of owning and developing intellectual property as the way forward for the company.

Generally, Company A feels that the Singapore government is pro-innovation and has put in place many initiatives that seek to promote greater innovation in the business community. Businesses in Singapore should be comforted that the legal protection regime currently in place in Singapore is robust and of world standard. However, Company A thinks that intellectual property awareness is still generally low in Singapore compared to other developed countries. The problem of counterfeits has not been completely eradicated, but ironically the presence of these counterfeits spurs Company A to innovate further and faster so as always to remain the forerunners in the competitive marketplace.

5.4.2 Company B – An Electronic Systems Company

Company B is one of the largest electronics system houses in the region. It is a wholly owned subsidiary of one of the largest publicly listed companies on the Singapore Stock Exchange. Company B's core business is in the design, implementation and delivery of innovative electronics system solutions for commercial, industrial, defence and public services applications worldwide. Company B was established in the 1960s and has since then become a leading provider of advanced electronics systems and solutions. Its business operations cover mainly the Asia-Pacific region and the company markets to more than 50 countries worldwide. In 2003, Company B's turnover was S$621 million, 30 percent of which was generated outside of Singapore.

Company B's expertise and innovative systems solutions are delivered through its three strategic business groups: (i) communication and sensor systems (ii) software systems (iii) large-scale systems.

Each of these strategic business groups has its own R&D units which undertake relentless pursuits in leading-edge technologies in their respective domains. R&D is a key component of Company B's strategy in the creation of cost-effective proprietary products and innovative system solutions for its customers. Company B's R&D efforts have resulted in the development of many innovative products, one of such innovative products playing a significant role in a recent national health crisis. Over the years, Company B's strong focus on R&D has won for itself numerous accolades and awards.

In the past few years, Company B has filed a total of 35 patents in Singapore and the other major markets such as the United States and the EU. At present, a total of 15 standard patents have been granted to Company B. Patent filing in countries outside of Singapore is done via the PCT.

In Company B, the decision of whether and how to file for a patent is undertaken with the participation of the senior management. During the company's monthly top management meetings involving all the general managers of strategic business groups, inventors present proposed filings for the meeting's approval. The meetings consider issues such as whether it is more advantageous from the company's perspective to file for patents or to keep the new technology as a trade secret, and if patent filing is the preferred mode of protection, which are the countries to file in. Since the filing process involves the general managers of all the strategic business groups, relevant IP-protection strategies and experiences are shared, and cross-fertilisation of ideas often also serves to enhance the inventions.

There are two aspects to Company B's IP strategy. One, the conventional defensive approach, is to use standard patents to stop competitors from copying their solutions and products. The other motivation for filing and owning patents is to use patents as a differentiating tool and to provide distinctiveness to Company B's systems solution offerings. The fact that Company B has proprietary ownerships such as patents in certain specific product features in the entire large system solution offered to its clients gives Company B a competitive edge against its competitors. This is because this puts the company in an advantageous position to solve the system problems of its clients in a unique way which is unrivalled by others. Patents are also important bargaining chips for Company B in the negotiation of alliances with business partners. At the present time, licensing revenues from the commercialisation of the patented technology are not one of the main objectives of Company B in its IP strategies. The oldest patent in Company B is currently five years old. In the electronics business, product lifespans are very short and hence Company B does not expect to maintain all its patents through their entire validity period. Besides patents, Company B also owns other intellectual property rights such as trade marks and copyrights.

What are Company B's views on the need for an alternative protection system for minor inventions and innovations such as the UM system in Singapore? Company B feels that it is always positive for the Singapore government to explore alternative protection systems with the view of encouraging greater creativity and innovation in businesses. As far as Company B is concerned, the UM system may be useful in protecting certain by-products of the company that have a heavy emphasis on design features. However, there is no immediate urgent need for such an alternative protection system in Company B.

With regard to the question of whether there is a need for Singapore to look into the possible implementation of an alternative protection system for minor inventions and innovations, Company B is of the opinion that the key issue is whether Singapore small and medium enterprises stand to benefit from such a model. To do so, the alternative system of protection must address the needs of such enterprises, and the SMEs must similarly be educated on the costs (including the value of the invention given away in a disclosure) and benefits of seeking protection under such a system.

Company B feels that on the whole the current patent system in Singapore has served its business objectives well. Since its inception in the 1990s, it has played a significant role in granting protection to many inventions and innovations in Singapore. It has put Singapore on the international map as a country with strong intellectual property protection laws. This in turns boosts confidence and accords certainty to businesses. However, it is noted that the current patent system in Singapore is relatively young and like any other system, it has its pitfalls. More time is therefore needed for the Singapore patent system to mature into a robust system that is at the same time responsive to businesses' needs.

5.4.3 Company C – A Water Treatment Company

Company C is a public-listed company on the Singapore Stock Exchange's Mainboard, supplying water treatment equipment and systems as well as its proprietary membranes. Company B's core business is in six main areas – seawater desalination; manufacturing process stream separation, concentration and purification; structured build-own-operate/transfer solutions; aquovators and home filter; and material science development in biodegradable plastics and cutting-edge technology in lactic acid production. Company C was established in 1989, and since then has grown into a leading regional player with market capitalisation of over S$480m, employing more than 480 staff in Singapore and China. In 2003, the company enjoyed a turnover of S$81.2m and turned in net profits of S$19.5m in 2003. About 50 percent of Company C's gross revenue earned in 2003 is generated overseas, particularly China.

Company C actively engages in R&D conducted in-house or in collaborations with leading institutions worldwide. The main areas of R&D are in membrane technology, biodegradable plastics and lactic acid production. Company C's relentless pursuit in developing leading-edge technology in these key areas of competence not only ensures that Company C stays ahead of its competitors in the field but also serves as a main thrust to put Company C on the world map as a leading water treatment company.

As an integrated solutions provider, Company C offers a comprehensive range of services which includes the design, fabrication, installation, instrumentation and maintenance of liquid treatment systems. Besides water treatment, Company C is also involved in the treatment of process industry fluids such as those used in the agri-food, pharmaceutical, biotech, chemicals and petrochemicals sectors. Company C has also made significant inroads into the food processing industry, in particular cane sugar process technology. Its stainless steel membrane separation technology, a technology developed by Company C's in-house R&D, has improved the yield, quality and efficiency of cane sugar production considerably at significantly lower costs.

To date, Company C has been granted a total of 10 patents filed in Singapore and in several other countries such as the United States, China and the EU. Patents filed overseas are made via the PCT which is more efficient and less costly. Presently, Company C is in the process of filing patents covering processes involving organic acids and other by-products. To strengthen and enhance its patent portfolio, Company C acquires patents of leading technology from other countries and also assists its business partners in the registration of patents in Singapore.

Company C possesses advanced technology for lactic acid production and it also owns one of the world's foremost patents on biodegradable plastics acquired through in-house research and jointly developed with reputable institutions worldwide.

The objectives of filing patents are manifold for Company C. It uses the patent system to protect its core competence and know how in the water treatment processes and products. Patents granted and acquired serve as an important notice to others in the market that Company C is in the business of water treatment and it has proprietary rights over the relevant processes and products. With a strong patent portfolio, Company C is seen as the leader in the relevant technology and this gives Company C a competitive edge when negotiating contracts with others. As a strategy, Company C prefers to maintain confidentiality of its new inventions until it is opportune to announce to the business community that it is in possession of such a technology. Therefore, not all new technology developed by in-house researchers or with collaborators that are patentable is automatically filed as patents. Company C makes careful decisions with regard to what technology to file and what to retain as collateral know how within the company. At the end of the day, business consideration dictates whether the patent system is the appropriate means of legal protection for the technology concerned. Similarly, business consideration is also the overriding factor whether a particular patent ought to be renewed or not and if it is to be renewed, what is the ultimate term of the patent.

Company C feels that patent registration increases its competitive edge, especially when the technology concerned is an important component of Company C's core competence. An example is Company C's ultrafiltration membrane technology. The proprietary rights over this technology have allowed Company C to compete advantageously in large-scale project tenders. The ability to manufacture its own membrane has resulted in substantial cost savings for Company C. Besides patents, Company C also has other intellectual property rights such as trademarks and trade secrets.

What are Company C's views on the need for an alternative protection system for minor invention and innovations such as the UM system in Singapore? Company C feels that in water treatment business which is a technology-intensive business, new inventions are rarely minor or petty in nature. Consequently, the demand for quick protection in the form of UMs in such line of business is likely to be low. Nevertheless, Company C is of the opinion that it is important to encourage innovation in Singapore, and any effort to encourage innovation by the government like the UM system must be carefully studied for its costs and benefits. It is also very important to provide alternative forms of protection to businesses. Thus, Company C agrees that there is a need for Singapore to look into the possible implementation of such an alternative system for minor inventions and innovations.

Company C feels that it has operated comfortably in the Singapore environment and it strongly believes that the intellectual property rights system of Singapore has in fact contributed to a relatively stable legal and business environment. It also notes that compared to a decade ago, Singapore now has more patent attorneys, examiners and scholars working in the field of intellectual property rights. Consequently, the level of service, quality of work, efficiency and expediency rendered by both the government and private sectors in the field of intellectual property rights to the business community has also improved considerably over the years.

5.5 CONCLUSION

Singapore's R&D expenditure and patenting activities have been increasing over the past decade, driven by large local and foreign enterprises. Patents have been widely used in the manufacturing sector across the semiconductor, ICT-related products and computer peripherals industries. Singapore remains in a stage of 'value adding' rather than 'value creation', but it is becoming a more creative economy. Yet, the quality of innovation indicates that the degree of product autonomy by sales was dominated by the manufacturing arm of parent companies and original equipment manufacturing rather than original brand design and design manufacturing. High-value-added innovation between 1991 and 2000 has improved in Singapore over time, as demonstrated by the rise in triadic family patents, which in relation to the country's GDP and population places Singapore on a level comparable to the developed world, albeit still substantially lower than the most advanced countries like Japan and the United States. Domestic ownership of foreign inventions has also risen, indicating that the Singaporeans are beginning to venture more abroad in terms of intellectual property filing while foreign ownership of local inventions has declined.

The UM system could stimulate innovation in the local SMEs sector while offering a low cost solution for its larger and foreign counterparts in Singapore for the following reasons:

- Less knowledge-based industries can also seek protection for those innovations meeting neither the necessary requirements nor the level of protection required in standard patents.
- High knowledge industries like the semiconductors, ICT-related products and computer peripherals can protect lower-end innovations with a more economical alternative.

However, it would be advisable to determine the long term sustainability of a UM system with a cost-benefit analysis, weighing carefully the pros and cons and their possible economic impact for which this study acts as a precursor. Moreover, there has to be a niche for the UM law to fill. One must identify what types of innovations are falling between the gaps of protection. It appears difficult to discern such subject matter in the Singapore context based on the questionnaire results. Indeed, the message is that industry feels there is a problem with complexity, length and cost of protection, rather than a need for yet another registration system.

Should Singapore introduce a new, cheaper and faster protection regime for a more macro-economic rationale i.e. to encourage innovation? The respondents in this study answered overwhelmingly in the affirmative with 95.2 percent stating that there was a need for a more rapid and cheaper protection system to encourage industrial growth and promotion. This poses a dilemma. On the one hand, industry is strongly of the view that a large public domain is necessary for incremental innovation – this presupposes a more lenient and relaxed protection regime, including the absence of a UM system or a weak UM system. On the other hand, industry is strongly of the view that there is no fast and cheap protection regime currently and that this may be discouraging industrial growth.

If incremental inventions are to be protected, there must be a valid reason as to why a UM system is to be introduced, and why this proposed law will institute a more favourable regime as compared with the current patent laws. After all, the government could alternatively extend design law, provide a misappropriation tort or more extensive passing off or unfair competition law, or simply improve the efficiency and 'user-friendliness' of the existing patent regime.

6. Australia

Andrew Christie and Sarah Moritz

Australia has seen two systems of second tier patents. The first such regime, the petty patent system, was introduced in 1979. Petty patents were replaced by innovation patents in 2001. The development of Australian second tier patent protection was a response to perceived deficiencies in patents and designs law. Prior to the introduction of petty patents it was thought that all inventions worthy of patent protection were able to achieve it. However, there was an apparent need for quick, less expensive and more easily obtainable patent protection for inventions with a short commercial life. Once petty patents were introduced, however, it became apparent that they not so much served inventors, as manufacturers. In light of this, and the recognition that existing designs legislation did not protect functional innovations, a gap was found to exist in the protection of minor and incremental innovations and inventions.[1] In 2001 a revised second tier patent was instigated – the innovation patent – which supplanted the petty patent.

The Australian experience of second tier patent protection provides a useful case study for countries contemplating the introduction of such protection. An analysis of both the petty and innovation patent systems illustrates possible models for second tier patent protection elsewhere. An analysis of second tier patents in the broader context of Australian patents, and a profile of the users of the Australian patent system indicate the effectiveness of petty and innovation patents in meeting the objectives for which they were introduced.

6.1 THE STANDARD PATENT SYSTEM

Prior to the Australian introduction of a second tier patent system, the relevant legislation concerning patents was the *Patents Act 1952* (Cth). In order for an invention to be patentable subject matter it was not allowed to have been published prior to the patent application being made, and needed to be new. The

[1] 5 ACIP, above n. 4, 23.

current legislation governing patents is the *Patents Act 1990* (Cth). Now the subject matter of a patent needs to be a manner of new manufacture in accordance with the *Statute of Monopolies*, be novel, non-obvious (i.e. evidence an inventive step over the prior art), useful and not have been secretly used. A patent grants 'the patentee the exclusive rights, during the term of the patent, to exploit the invention and to authorise another person to exploit the invention'.[2] The term of a standard patent is twenty years. The *Intellectual Property Laws Amendment Act 1998* (Cth) made it possible for patents concerning pharmaceutical substances to receive an extended term for up to five years. The term of a standard patent runs from the date of lodgement of the specification.

6.2 PETTY PATENTS – AUSTRALIA'S INITIAL SECOND TIER PATENTS

6.2.1 History of the Petty Patent

In Australia, second tier patent protection was largely initiated by the findings of the Designs Law Review Committee (the Franki Committee) in its *Report Relating to Utility Models*. Around the time of the review, the Australian Patent Office had been under considerable pressure due to increasing patent application numbers. Australia followed the German system, and in 1969 the *Patents Act 1952* (Cth) was amended to introduce a system of deferred examination, where examination occurred upon request of the applicant, but no later than five years after a patent application was made. The idea was put forward that UM protection without substantive examination, such as existed in Germany, might be appropriate for the Australian situation.[3] The Franki Committee assessed whether there existed:

> a need for a form of protection for lesser technological developments which have merit, but are not entitled to protection under the designs legislation and are not patentable because they do not involve a sufficient inventive step.

The committee recommended the introduction of a system of petty patents based on the following findings:

---

[2] s13(1) of the *Patents Act 1990* (Cth). In the dictionary of the Act 'exploit' is defined as 'make, hire, sell or otherwise dispose of the product, offer to make, sell, hire or otherwise dispose of it, use or import it, or keep it for the purpose of doing any of those things; or where the invention is a method or process – use the method or process or do any act mentioned above in respect of a product resulting from such use'.

[3] Franki Committee, above n. 9, 7.

- There was no gap in the industrial protection available in Australia – all inventions which merited protection could achieve it.[4]
- There was a need for quick and less expensive protection which could be easily obtained for inventions with a short commercial life.[5]
- Design registration was inadequate, as design registration merely protected the aesthetics and not the functionality of a design.[6]
- Standard patent protection did not adequately cater for inventions of a shorter commercial life.
- The deferred examination procedures for standard patent applications were insufficient for many inventors, as inventors often required a patent before a manufacturer would consider investing in its production.
- The grant of patent rights for smaller inventions was too slow, too expensive and too difficult, with the result that the inventors were either unable to exploit their inventions commercially themselves or to interest manufacturers in doing so.

The possibility of including provisions for additional protection in the designs legislation was discussed but rejected because it was thought that the purpose of designs legislation was of a different nature, namely the protection of features of shape, configuration, pattern or ornament applicable to articles and not for inventions. Secondly, it was considered that additional protection could be introduced by offering protection against copying only. However, it was thought that monopoly rights were appropriate in this instance, rather than protection from copying.

6.2.2 Objectives of the Petty Patent

The objectives of the petty patent system were to create a form of patent protection:

- that was less expensive, easier and more quickly obtained than standard patent protection;
- that would be used for inventions with a relatively short commercial life[7];

---

[4] This was similar to the negative finding of the Banks Committee, which assessed whether second tier patent protection should be introduced in the United Kingdom.

[5] T.D. Mandeville Report et al., Economic Effects of the Australian Patent System – A Commissioned Report to the Industrial Property Advisory Committee (1982), 11.

[6] For a discussion on the inadequacies of design registration for the protection of inventions see also Australian Law Reform Commission Report (1994), 'Designs', discussion paper no. 58 and (1995), report no. 74.

[7] Designs Law Review Committee ('Franki Committee'), Parliament of Australia

- primarily intended for Australian industry and inventors.[8]

It was thought that this would encourage inventors and SMEs to seek patent protection even if the subject of their patents was only likely to reap rewards for a relatively short time period. Hence, inventors and SMEs were expected to receive a greater return on their investment, and be encouraged to undertake further innovation. As the monopoly rights granted to a petty patent holder were more readily achieved than through standard patents, the term of a petty patent was shorter than for a standard patent.

The principal means by which the petty patent system aimed to achieve its objectives was through:

- eliminating opposition proceedings prior to grant which would increase the speed of an application being granted and reduce expenses associated with opposition hearings;
- restricting the petty patent to one claim which would reduce patent attorney costs.

6.2.3 Features of the Petty Patent

The petty patent regime was operative from July 1979 to May 2001. Unlike many forms of second tier patent protection, the subject matter for which petty patents were permitted was identical to standard patents.[9] The requirements of patentability were also initially identical.

Petty patents received an initial one year term of protection from the date of sealing, with a maximum term of six years from the date on which the patent application was lodged. During the initial year of a petty patent, evidence pertaining to grounds of invalidity could be brought by third parties to the commissioner. It was thought that this would increase the presumption of validity, especially given the lesser examination procedures of petty patents compared with standard patents. Provisional applications could not be made for petty patents until the *Petty Patents Act 1990* (Cth) took effect. However, prior to this, a provisional

---

(1973), 'Report relating to utility models', Parliamentary report no. 121, 11. The view was that the time and cost associated with a standard patent meant that in practice there was 'not a sufficiently quick and inexpensive and simple means of providing protection for the lower range of inventions, especially small articles having short commercial lifespans'. Industrial Property Advisory Committee (IPAC) (1978), 'Report on proposed petty patents legislation', 3.

[8] IPAC (1981), *Petty Patent Seminar*, 1.

[9] For example, many utility models only cater for minor mechanical and electrical inventions, and accordingly have a lower level of inventiveness.

application could be lodged for a standard patent, and once a complete standard patent application was made, a divisional could be made from that application for a petty patent. In this way the earlier priority date, based on the provisional application, could be claimed for a petty patent. At its introduction the petty patent regime only allowed for one claim in the petty patent specification.

There were no opposition proceedings prior to the grant of a petty patent, and the examination process was intended to be limited in that the Commissioner of Patents had only to 'consider' whether there existed any lawful grounds for objection to the petty patent.[10]

Prior to the grant of a petty patent, a petty patent application could be converted to a standard patent application. Given the intended restricted examination procedures, the time in which such a conversion could occur was limited. Like standard patents, petty patents could be applied for via the Patent Cooperation Treaty route. However, as petty patents were generally intended for the exploitation of inventions specifically within Australia, international protection through the PCT route was generally of little interest.[11]

6.2.4 Review of the Petty Patent System

In 1981, an assessment was made to assess the success of the petty patent system. The primary advantage of the petty patent system was the speed with which protection was obtained, with 90 percent of petty patents being granted within three months of being filed. The main criticisms were as follows:

- The petty patent system did not serve the people for whom it was originally intended.
- The cost of a petty patent was comparable to standard patents, as utmost care was required when drafting the single claim upon which the validity of the petty patent depended.
- The single claim also made it difficult to enforce a petty patent: if the claim was too broad the patent would be invalid for obviousness or want of novelty, and if it was too narrow a potential infringer could escape liability for infringement.

---

10 In practice this amounted to an examination of the application, even though it appears it was not the intention of the legislature that it be so: see Ricketson, Staniforth (1984), *The Law of Intellectual Property*, [47.56].

11 According to World Intellectual Property Organization data, throughout the duration of the petty patent system only 13 petty patents were granted, which had been applied for via the PCT route (see World Intellectual Property Organization (2004), '25 years of industrial property statistics (1975–2000)', accessed 20 March, 2004 at www.wipo.int/ipstats/en/publications/25_years/index.htm.

- The six year term of petty patents was criticised for being too short to truly provide an incentive for a potential manufacturer to invest.

A subsequent economic review of the Australian patent system, *Patents, Innovation and Competition in Australia,* recommended change, but continued use of the patent system. The report found that while the economic effects of the patent system may be modest, the transition costs of withdrawal from the international system might be much larger.[12] Several recommendations were made which were subsequently incorporated in the *Patents Act 1990* (Cth), operative from 1 May 1991:

- A two-tier system of patent protection with differing standards of novelty and obviousness was introduced: the report recommended that standard patent applications be assessed against the prior art base of the world, compared with an Australian prior art base for petty patents.
- The number of permissible claims of a petty patent was increased to a maximum of one independent claim with two dependent claims.
- One invention could not be the subject of a standard patent and a petty patent.

However, the government's Science and Engineering Council (PMSEC) found that the petty patent system did not provide 'limited term protection for lower levels of invention such as small-scale incremental advances in existing or newly adapted technologies'.[13] The council suggested the two main reasons for this were the absence of differentiation between the level of inventiveness required of standard and petty patents, and the relatively short term (six years) for which petty patents were granted. The Advisory Council of Intellectual Property released its *Review of the Petty Patent System* which looked at the issue of whether functional innovations received adequate protection through the existing standard and petty patent systems. This review found the petty patent system was not providing protection for incremental innovations. It wrote:

> [t]he 'gap' relates to functional innovations that are not sufficiently inventive under the present standard or petty patent system to warrant protection, and are not protect-

---

[12] IPAC (1984), *Patents, Innovation and Competition in Australia*, 1. Another report into the economic implications of the Australian patent system, however, found the benefits of the patent system to be too tenuous and subtle and the overall benefit/cost ratio negative, such that there was no economic justification for extending patent monopolies. T.D. Mandeville et al. (1982), 'Economic effects of the Australian patent system – a commissioned report to the industrial property advisory committee', 213.

[13] Prime Minister's Science and Engineering Council (PMSEC) (1993), *The Role of Intellectual Property in Innovation*, vol. II, p. 70.

> able under the designs system which protects the appearance of articles, but not 'the way they work'. Provision of protection for these incremental innovations will encourage Australian individuals and businesses to invest in the development and marketing of their 'good ideas' in the domestic market.[14]

Accordingly, ACIP recommended that the existing second tier patent system be changed to provide fast, limited monopoly protection for lower-level or incremental inventions. This new 'innovation' patent system would require a lesser level of inventiveness than petty patents.

6.3 INNOVATION PATENTS – AUSTRALIA'S SUBSEQUENT SECOND TIER PATENTS

6.3.1 Objectives of the Innovation Patent

The innovation patent system was introduced in July 2001, its primary objective being to fill the 'gap' that existed with regard to minor and incremental innovations. Secondly, it offers a quick, less expensive and simple form of protection to encourage individuals and small to medium sized businesses to realise their good ideas.[15] In addition, the inventiveness threshold for innovation patents was reduced to render a greater range of lesser innovations patentable.

It is thought that the innovation patent will particularly assist SMEs obtain patent protection, and protection for a sufficiently long period to encourage investment in innovation.

6.3.2 Features of the Innovation Patent

To be a patentable invention for the purposes of an innovation patent an invention must involve an innovative step[16] as opposed to the inventive step required of standard patents. An invention is understood to evidence an innovative step 'unless the invention would, to a person skilled in the relevant art, in the light of the common general knowledge as it existed in the patent area before the priority date of the relevant claim, only vary from the kinds of information set out in subsection (5) in ways that make no substantial contribution to the working of the invention'.[17]

The main features of innovation patents are:

---

14 Advisory Council of Intellectual Property, *Review of the Petty Patent System*, 5.
15 Janis, *ibid.*, 171.
16 s18(1A)(b)(ii) *Patents Act 1990* (Cth).
17 s7(4) *Patents Act 1990* (Cth).

- patentable subject matter for the purpose of innovation patents is the subject matter for which standard patent protection is available, with the exception of inventions concerning plants and animals and processes for the generation of plants and animals. Inventions concerning plants and animals are, however, capable of innovation patent protection where such inventions are microbiological processes or products of such processes;
- a maximum eight year term of protection;
- the possibility of five claims in the specification;
- they are not subject to opposition proceedings prior to grant;
- they are only subject to substantive examination at the direction of the commissioner, or after grant at the request of the patentee or a third party;
- the prior art base applicable to innovation patents is that of standard patents; however, the threshold level of inventiveness is lower;
- the innovation patents cannot be filed through the PCT route.

The prior art base, which is applicable to standard patents and to innovation patents, comprises documents publicly available anywhere in the world, as well as acts which have taken place anywhere in the world.

6.4 EMPIRICAL ANALYSIS OF AUSTRALIAN SECOND TIER PATENT SYSTEMS

6.4.1 Introduction

This section depicts the use of second tier patent systems in the Australian context on the basis of empirical data. Firstly, second tier patents are situated in the context of the broader Australian patent system. Secondly, a profile of the users and uses of the Australian second tier patent systems is established. Data on the use of the petty and innovation patent systems is analysed in detail according to applicant type, country of origin and technology type of the claimed invention, and this information is compared with data on the use of standard patents. This analysis establishes that the profile of the users of the standard patent system differs from the profile of the users of the second tier patent system.

Finally, petty patents were introduced in Australia because of the perceived gap at the interface of patent and design protection for innovations of short commercial value. The average success rate for design registration applications between 1985 and 2003 was 81 percent.

6.4.2 Second Tier Patent Systems in the Australian Patent Context

Upon comparing second tier patent application numbers to standard patent application numbers, it is found that both second tier and standard patent applications have generally increased with time (Figure 6.1). Further, although second tier patent applications account for only a small portion of total patent applications, this portion is by no means negligible and is indeed increasing.

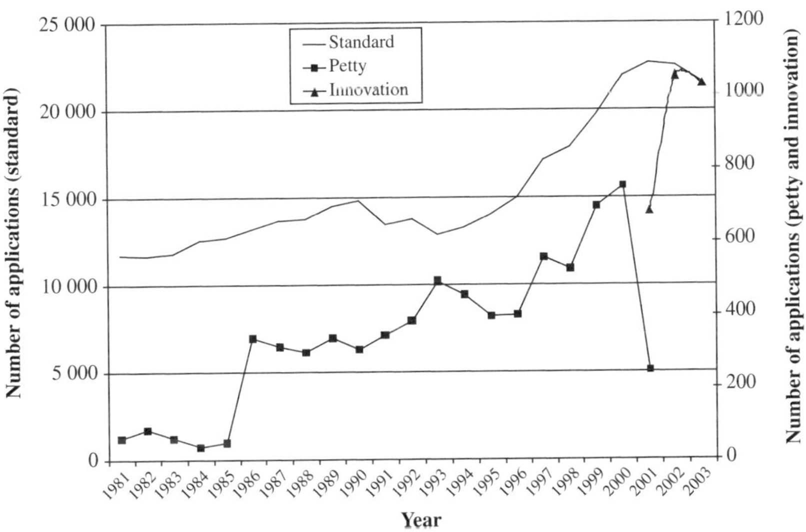

Source: IP Australia data.

Figure 6.1 Standard (1981–2003), petty (1981–2001) and innovation (2001–2003) patent applications

Petty patents came into existence in 1979. Since then there has been a general increase in standard and total patent applications, with a small decrease in standard patent applications between 1990 and 1993, possibly due to economic conditions at the time. The most dramatic increase in total patent applications occurred between 1993 and 2000. This corresponds to an increase in patenting throughout the world, which occurred in the past decade.[18] Since 1986 the

---

[18] In 2002 more than 850 000 patent applications were filed in Europe, the United States and Japan. This represented an increase of 250 000 since 1992 (see OECD (2004)), *Patents and Innovation: Trends and Policy Challenges*, 7, accessed 20 March, 2004 at www.oecd.org/dataoecd/48/12/24508541.pdf.

number of second tier applications has generally increased. Recently there has been a decline in relation to standard patent applications. In 2001, 22 742 entered the Australian patent system. By 2003 this number had declined to 21 623. When one accounts for the time lag in PCT patent applications entering the national phase, the actual decline in innovative activity associated with the decrease in standard patent applications occurred prior to 2000.

At the same time as standard patent applications decreased, innovation patent applications increased, which has led to second tier patents constituting a greater proportion of total patent applications. The transition from the petty patent system to the innovation patent system in 2001 has been associated with a significant increase in second tier patent applications. From 1998 to 2002 second tier patent applications more than doubled. As a result, innovation patents now account for a larger portion of total patent applications than did petty patents. Whereas in 1993 petty patents constituted 3 percent of total patent applications, in 2003 innovation patent applications constituted 5 percent of total patent applications.

6.4.3 Individual Versus Company Applicants

Standard patents

From 1979 to 1987 the proportion of individual applications for standard patents generally increased, after which point it stabilised until 1994 (Figure 6.2). At this time applications for standard patents made by individuals constituted 16–17 percent of all applications. From 1994 to 2001 this proportion has been decreasing. In 2001 individual applications were only 13 percent of total applications. This decline in standard patent applications made by individuals has coincided with a rapid increase in PCT applications and gradual decrease in non-PCT applications. This is not surprising as there is a greater representation of companies among PCT applications than among non-PCT applications.

Petty patents

From 1979, the proportion of individual petty patent applications generally increased, until 1989, after which this proportion by and large decreased (Figure 6.3). In 1989 individual applications constituted 84 percent of total petty patent applications, whereas by 2001 the proportion of petty patent applications made by individuals was only 51 percent.

Innovation patents

In 2001, individual applications constituted 66 percent of total innovation patent applications, whereas by 2003 the proportion of innovation patent applications made by individuals increased by three percentage points to 68 percent (Figure 6.4). Whether this is indicative of a general trend to come is difficult to ascertain

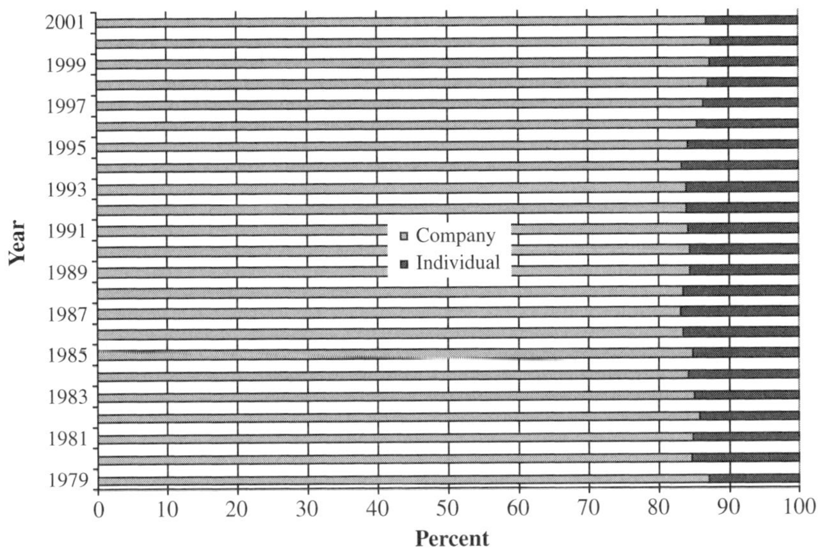

Source: IP Australia data.

Figure 6.2 Company versus individual standard patent applications

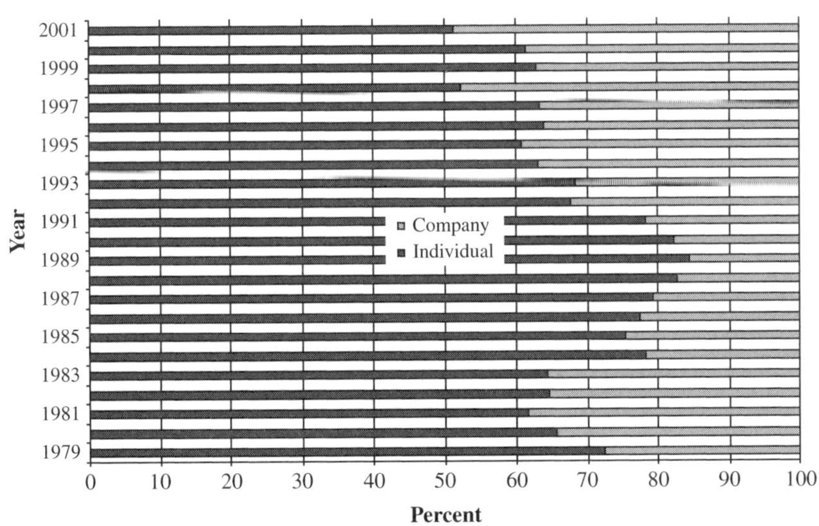

Source: IP Australia data.

Figure 6.3 Company versus individual petty patent applications

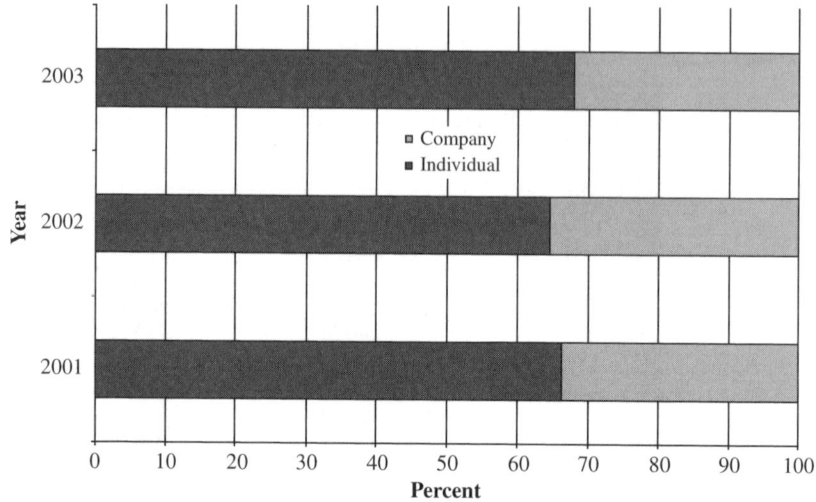

Source: IP Australia data.

Figure 6.4 Company versus individual innovation patent applications

given that the innovation patent system has only existed for such a short period.

Average proportion of individual versus company applications

Averaged from 1979 to 2001, the proportion of standard patent applications made by a company was 85 percent, and the remaining 15 percent of applications were made by individuals (Figure 6.5). From 1979 to 2001, the average proportion of petty patent applications made by a company was 31 percent of total petty patent applications. The remaining 69 percent of applications were made by individuals. This contrasts markedly with standard patent applications. The proportion of total petty patent applications made by individuals is more than four times greater than this proportion in relation to standard patent applications. Averaged from 2001 to 2003, 66 percent of innovation patent applications were made by individuals and 34 percent of applications were made by companies.

6.4.4 Australian Versus Foreign Applicants

Standard patents

Australian applications are decreasing as a proportion of total standard applications (Figure 6.6). In 1994, Australian applications accounted for 14 percent of

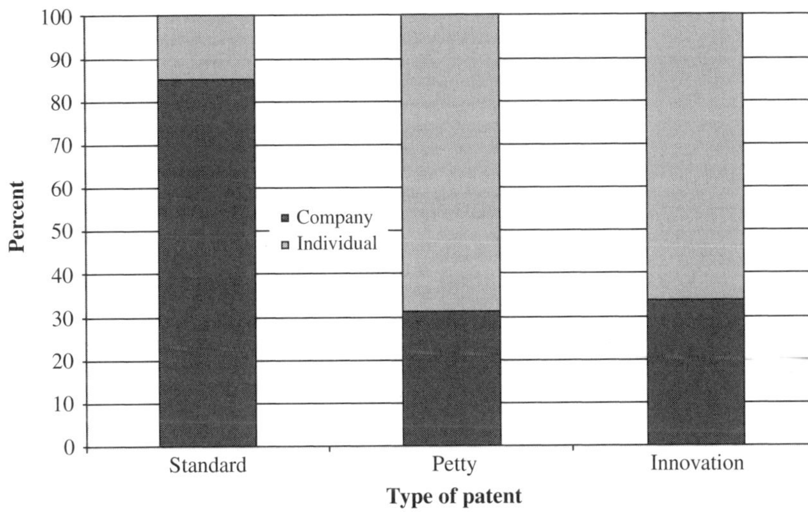

Source: IP Australia data.

Figure 6.5 Average proportion of individual versus company patent applications: standard and petty patents (1979–2001), innovation patents (2001–2003)

total standard patent applications. However, this proportion has been in steady decline as the number of overseas PCT national phase entries has increased. In 2001, Australian applications made up only 10 percent of total standard patent applications.

Petty patents

Although the majority of petty patents applications from 1979 to 2001 were Australian, this proportion generally decreased from 1983 (Figure 6.7). Then, Australian applications accounted for 93 percent of total applications. When the petty patent system concluded in 2001, Australian applications represented a significantly reduced 79 percent of total petty patent applications.

Innovation patents

The innovation patent system is predominantly utilised by Australian companies and individuals (Figure 6.8). Further, at the inception of the innovation patent system, Australian applications dominated total innovation patent applications with a greater majority than did Australian applications at the conclusion of the petty patent system (87 percent as opposed to 79 percent).

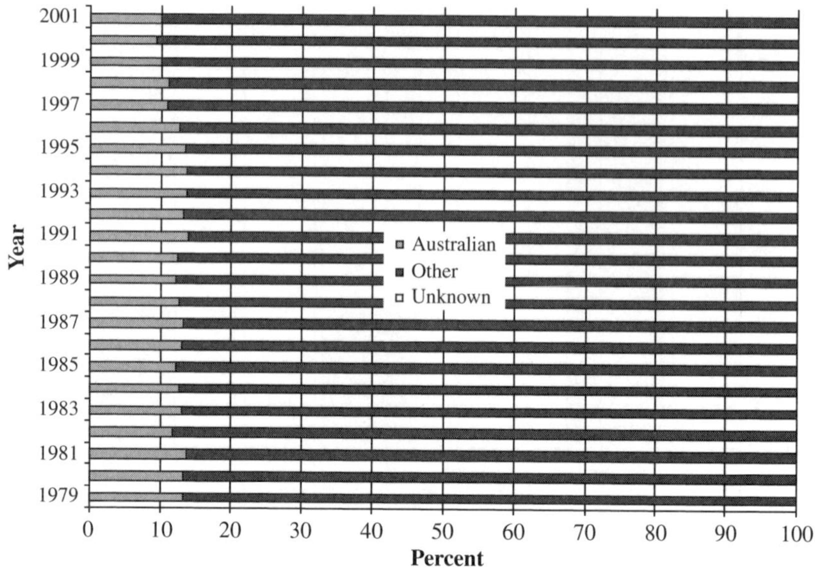

Source: IP Australia data.

Figure 6.6 Australian versus foreign standard patent applicants

Average proportion of Australian versus foreign applicants

From 1979 to 2001 Australian standard patent applications accounted for 12 percent of total standard patent applications, and foreign applications made up 88 percent of this total (Figure 6.9). In stark contrast to applications for standard patents, the average proportion of Australian to foreign petty patent applications during this period was 81 percent to 13 percent. The remaining 6 percent cannot be classified as either Australian or foreign, as the country of origin is unknown. Essentially, as far as petty patent applications are concerned, the proportion of Australian versus foreign patent applicants is the inverse of standard patents. The average proportion of Australian applications for innovation patents is relatively stable and greater than for petty patents. Australian applications for innovation patents account for 87 percent of total applications, with the remainder coming from other countries.

6.4.5 Country of Origin of Foreign Applicants

Standard patents

Averaged from 1979 to 2001, foreign standard patent applications accounted for 88 percent of total standard patent applications (Figure 6.10). The top five

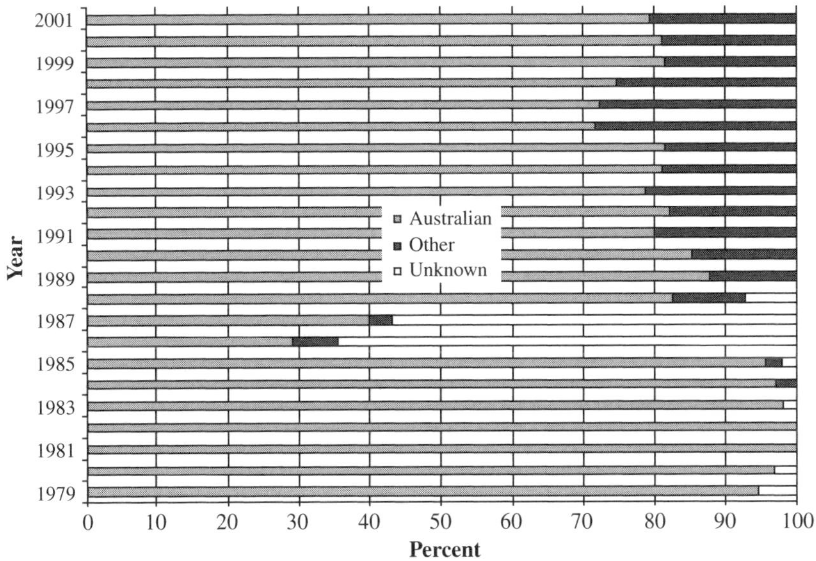

Source: IP Australia data.

Figure 6.7 Australian versus foreign petty patent applicants

foreign countries from which standard patents were applied during this period were:

- United States
- Japan
- United Kingdom
- Germany
- France.

Petty patents

Averaged from 1979 to 2001, foreign petty patent applications accounted for just 13 percent of total petty patent applications. The top five foreign countries from which petty patents were applied during this period were (see Figure 6.11):

- Taiwan
- United States
- New Zealand
- United Kingdom

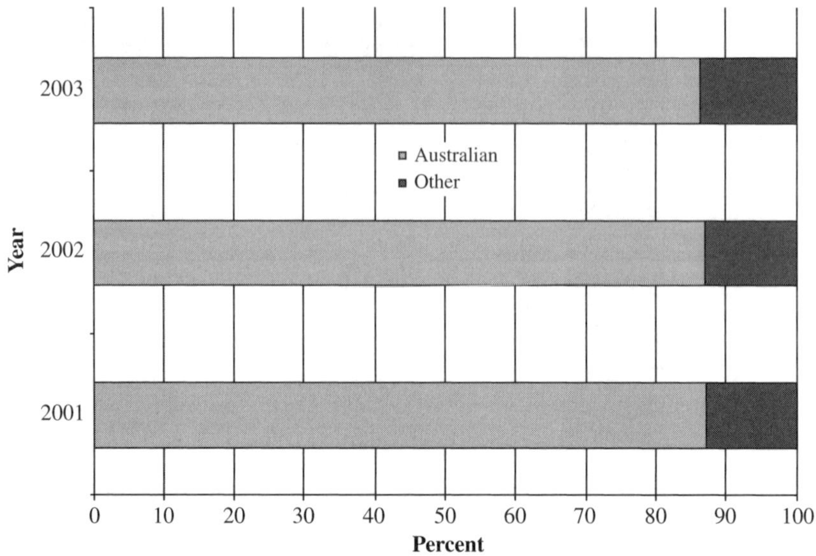

Source: IP Australia data.

Figure 6.8 Australian v foreign innovation patent applicants

- Germany.

Compared with the top five foreign countries for standard patent applications in Australia, Japan and France are not represented, and instead Taiwan and New Zealand are, and Taiwan is even positioned before the United States. In total, applications for petty patent applications from the top five foreign countries account for 60 percent of all foreign petty patent applications.

Innovation patents

Averaged from 2001 to 2003, foreign innovation patent applications accounted for just 13 percent of total innovation patent applications (Figure 6.12). The top five foreign countries from which innovation patents were applied during this period are:

- Taiwan
- United States
- New Zealand
- United Kingdom
- China.

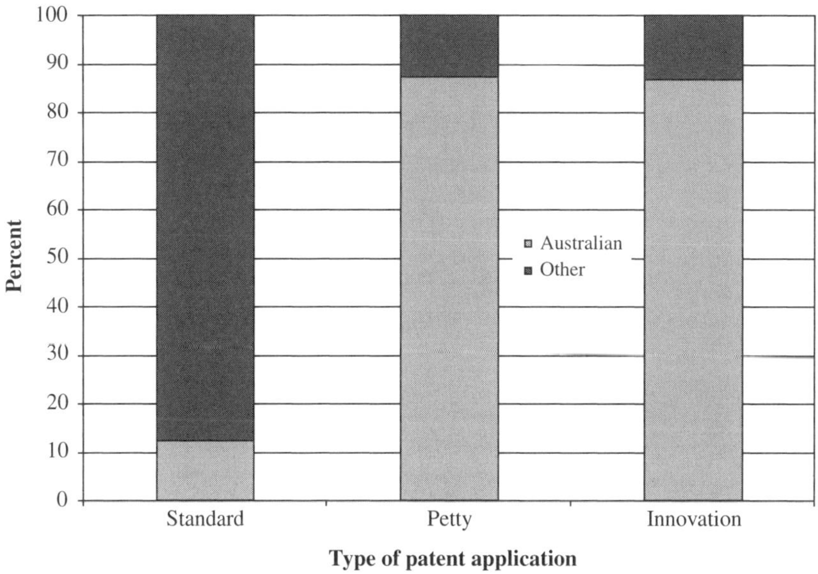

Source: IP Australia data.

Figure 6.9 Average proportion of Australian versus foreign patent applications: standard and petty patents (1979–2001) and innovation patents (2001–2003)

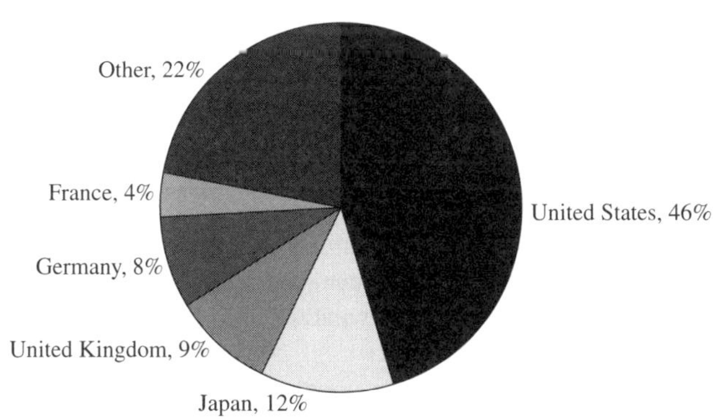

Source: IP Australia data.

Figure 6.10 Average top five foreign countries for standard patent applications

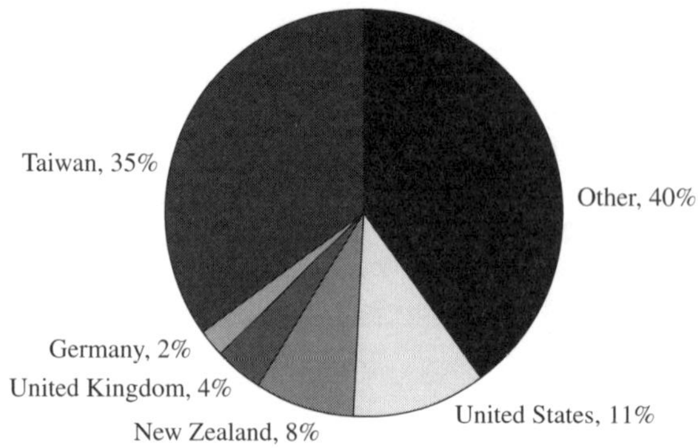

Source: IP Australia data.

Figure 6.11 Average top five foreign countries for petty patent applications

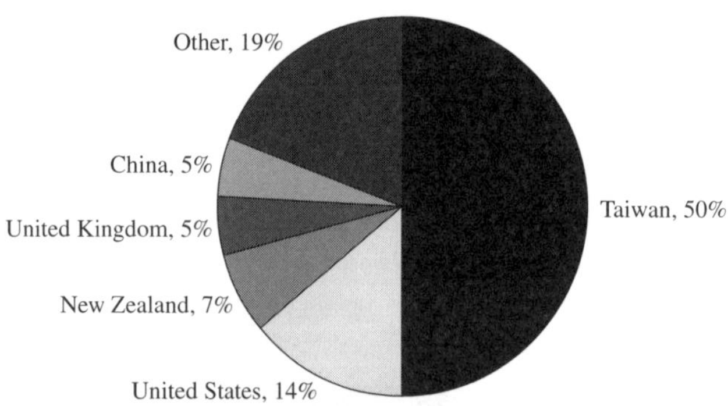

Source: IP Australia data.

Figure 6.12 Average top five foreign countries for innovation patent applications

The five foreign countries from which most foreign innovation patent applications originate are the same as for petty patent applications, except that in fifth place Germany is replaced by China. In total, applications for innovation patent applications from the top five foreign countries account for 81 percent of all foreign petty patent applications.

6.4.6 Technology Groups of Applications

Patent applications are classified according to the Office of Science and Technology (OST) classification through which each particular International Patent Classification (IPC) subclass is mapped to a more general OST technology group. The average representation of each technology group as a proportion of total standard, petty or innovation patent applications over the given time period was calculated. The five most represented technology groups for each patent type were then compared.

Standard patents

The top five technology groups for standard patents are all knowledge-intensive areas, which is perhaps not surprising given the high level of inventiveness necessary in order to receive standard patent protection (Figure 6.13). These technology groups are:

- organic fine chemicals
- pharmaceuticals, cosmetics
- medical engineering
- telecommunications
- analysis, measurement, control.

Source: IP Australia data.

Figure 6.13 Average top five technology groups for standard patents

Petty patents

None of the technology areas for which standard patents are predominantly sought appeared within the top five technology groups for petty patents. Instead,

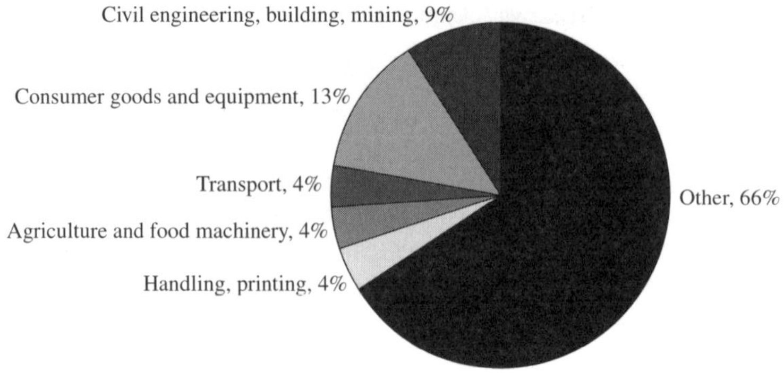

Source: IP Australia data.

Figure 6.14 Average top five technology groups for petty patent applications

the five most represented technology groups for petty patent applications were (Figure 6.14):

- consumer goods and equipment
- civil engineering, building, mining
- handling, printing
- agriculture and food machinery
- transport.

Innovation patents

Applications for innovation patents are largely made in relation to the same technology groups as petty patents, albeit in different proportions. The most commonly represented technology groups for innovation patents from 2001 to 2003 were (Figure 6.15):

- consumer goods and equipment
- civil engineering, building, mining
- transport
- information technology
- handling, printing.

Whereas the top five technology groups account for 30 percent and 34 percent of standard and petty patent applications respectively, they account for 59 percent of innovation patent applications.

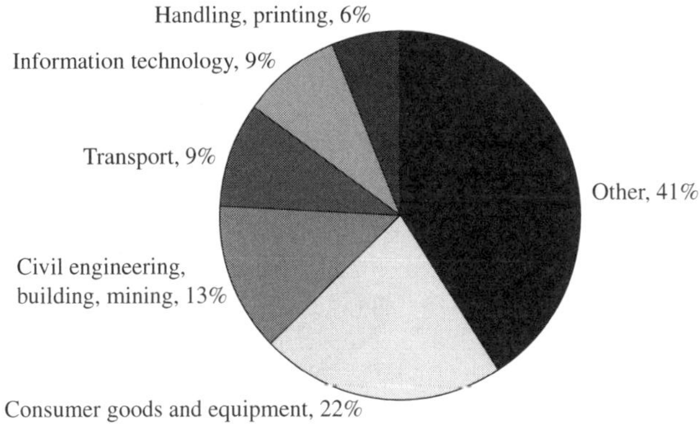

Source: IP Australia data.

Figure 6.15 Average top five technology groups for innovation patents

6.5 CONCLUSION

Three basic findings may be put forward. First, both petty and innovation patents were intended to cater specifically for individual inventors and SMEs. A breakdown of applications according to individual and company applicants suggests that petty and innovation patents have indeed appealed specifically to individuals. Second, an analysis of whether applicants for patents are Australian or foreign suggests that, relative to standard patent applications, second tier patents are utilised more by domestic inventors and innovators, for whom they were primarily intended, than foreign inventors. Third, an analysis of the data according to technology group illustrates that second tier patent systems in Australia have been and continue to be used in relation to technology types other than those in relation to which standard patents are sought.

Petty patents were introduced to provide less expensive protection for inventions of short commercial value, in particular for inventors and SMEs. It was thought that a less expensive form of patent protection would assist inventors, because patent protection would be more accessible to them. In turn, inventors would be placed in a better position to enter into contracts with manufacturers for the commercialisation of their inventions. Accordingly, one would anticipate greater representation of individuals among petty patent applicants than standard patent applicants, and this was the case.

The petty patent was, however, criticised during the course of its existence for not adequately serving those for whom it was initially introduced. As just stated, short term protection was of little value to potential manufacturers, and the single claim became so important that drafting it was as costly a task as drafting a complete standard patent specification. Perhaps these problems with the system played a part in the declining representation of individuals among petty patent applicants. By the time of its cessation in 2001 only 50 percent of petty patent applications were made by individuals. A greater proportion of innovation patent applications is made by individuals than was the case in the latter years of the petty patent system. As previously mentioned, innovation patents were intended to assist individual inventors and SMEs by lowering the inventive threshold requirement in order to allow incremental and minor innovations to receive patent protection.

Generally there exists an inverse relationship between Australian and foreign standard and second tier patent applications. The proportion of foreign standard patent applications approximates the proportion of Australian innovation patent applications. Averaged from 1979 to 2001, Australian applicants constituted 81 percent of total petty patent applicants. Similarly, the average proportion of Australian innovation patent applicants from 2001 to 2003 was 87 percent. It would appear that the characteristics of petty and innovation patents appeal to domestic innovators, these characteristics being the less expensive, more quickly obtained patent protection of shorter duration and, with regard to innovation patents, the lesser requirement of inventiveness. It is also likely that data concerning second tier patent applications relates to innovations for which the party applying is not seeking protection overseas, and in the majority of cases innovators who exclusively seek protection within the Australian market will be Australian innovators. Assuming that the majority of foreign applicants for Australian patents also seek to profit from their inventions in other countries, they will be likely to use the PCT application route. Although the PCT application route was available for petty patents, the delay involved apparently negated the positive effect of the quicker grant process – the primary advantage of petty patents. Only very few PCT applications were made for petty patents. PCT applications cannot be made for innovation patents.

There is no commonality between the top five technology groups for which standard patent applications are made and the top five technology groups for which petty and innovation patent applications are made. Petty patents were applied for in relation to technologies other than those for which standard patents were sought, and for virtually the same technology groups as innovation patents. This would suggest that the differing threshold of inventiveness for innovation patents has little, or no, bearing on the subject matter for which second tier patents are sought. Certain specific characteristics of second tier patents, such as the speed with which they are granted and the lesser cost, appear to suit specific technology areas, and these technology areas are similarly represented among innovation patent applications. It would be relevant to know whether inventions within the top five technology groups for petty patents, such as the category of consumer goods and equipment, generally have a shorter commercial life than inventions for which standard patent protection is sought. One might speculate that inventions of this category tend to be 'fad' inventions, which have only limited commercial value, in which case the primary objective of the petty patent system might to some extent have been met.

In relation to innovation patents it would be necessary to establish whether the technology groups for which innovation patents are most commonly sought, for example consumer goods and equipment, tend to be areas in which innovations are incremental. The presence of information technology among the top five technology groups for innovation patents is, perhaps, because this is an area in which the quick grant of a patent is particularly relevant, in order to prevent competitors from copying processes as soon as possible. Developments in information technology occur at such a rapid speed that the value of the quick protection offered by innovation patents is arguably heightened.

The analysis of the extent to which the objectives of the petty and innovation patent systems have been met suggests that the innovation patent system best meets its objectives. Indeed, innovation patents appear to be serving the purposes of *both* the petty and innovation patent systems.

7. Japan and South Korea

Graham Dutfield and Uma Suthersanen

7.1 ECONOMIC AND INNOVATION CLIMATE IN JAPAN

Does an expansion of the scope of patent rights induce more innovative effort? An analysis was undertaken in 2000 gauging whether the 1987 Japanese patent reforms had had any impact. The reforms had expanded the scope of patent: however, the authors found *no* evidence of a statistically or economically significant increase in either R&D spending or innovative output that could plausibly be attributed to these reforms.[1]

In a more recent 2004 report, Japan ranks ninth in terms of competitiveness though much of its R&D spending is by large corporations, rather than SMEs.[2] Nevertheless, another report states that Japan languishes in the middle of the OECD nations in terms of investment in knowledge creation, R&D by SMEs and venture capital investment as a percentage of GDP; the same report gloomily states that as Japan's technology gap with the United States widens, its edge over China is shrinking.[3] Does the shrinking SME contribution explain why Japanese UM registration statistics are falling?

7.2 HISTORICAL BACKGROUND

The first patent law was the 'Patent Monopoly Act' which was proclaimed publicly in 1885. In addition to this, the UM law was enacted shortly after in 1905, in order to complement the patent system. The main aim of the UM system was to protect less significant inventions and to foster technologies of small

[1] Sakakibara, M., L. Branstetter and M.P. Page (2001), 'Do stronger patents induce more innovation? Evidence from the 1988 Japanese patent law reforms', *Rand Journal of Economics*, accessed at www2.gsb.columbia.edu/faculty/lbranstetter/PDFpapers/Do_stronger_patents_induce_more.pdf.

[2] World Competitiveness Yearbook 2004, IMD.

[3] Development Bank of Japan (2005), research report no. 53, 'Japan's innovative capacity and policies for commercialising new technologies'.

and medium sized enterprises. The UM system was thought to be rapid and easier, and thus particularly suitable for SMEs. By the end of the 1980s, however, there were long delays in the grant of UM applications partly because the cost difference between the patent and UM systems is negligible, and partly because of an increase in the number of applications for UMs by both small and large companies. In response, the Japanese government revised the UM law to allow accelerated registration without substantive examination, and to shorten the term of protection to six years from the filing date on the basis that product lifecycles were expected to shorten. The law now grants a right on the basis of a judgement of whether the application satisfies a series of requirements (referred to as the basic requirements) judged to be necessary for registration. The current applicable legislation is the Utility Model Law No. 123 of 13 April 1959, as amended by Law No. 220 of 22 December 1999.

7.3 JAPAN'S UTILITY MODEL SYSTEM

Subject Matter and Criteria of Protection

The Japanese UM system grants protection to devices that are defined as 'the creation of technical ideas by which a law of nature is utilised'. Consequently, ideas or methods relating to products are not protectable subjects. Such devices must relate to shapes of items, their structure or combination. Devices which are liable to contravene public order, morality or public health are statutorily excluded from protection.

The Japanese law requires the device to be novel, inventive and industrially applicable. Novelty is tested by reference to global use or publication anywhere in the world. A device will not satisfy the criterion of inventiveness if it could very easily have been made prior to the filing of the UM application by a person with ordinary skill in the art to which the device pertains. However, a high level of creativity as applicable under patent law is not required for UMs. The law does not provide a definition of what industrially applicable means.

Procedure and Examination

The first to file principle under the UM law differs from the patent law. If two or more UM applications are filed on different dates relating to the same device, the earlier application filed shall obtain a registration. If they are filed on the same date, no one shall obtain a registration. This presupposes that there shall never be a mutual consultation among the applicants to agree on a sole applicant among them – which is unlike the situation under patent law. However, when an application has been abandoned, withdrawn or dismissed, the registration shall be granted to the other application that was filed on the same date as the withdrawn, abandoned or dismissed application.

Upon receipt of the application the patent office conducts an initial examination to check if the application conforms to the basic requirements of formality – there is no detailed examination as to the substantive aspects of the application unlike in patent applications. Basic requirements are as follows:

- The applied-for invention relates to the shape or construction of articles or a combination of articles.
- The claimed invention is not liable to contravene public order and morality.
- The application satisfies the requirements for the format of the claims as well as the unity of application.
- All the necessary items are described in the specifications and drawings, and in addition these descriptions are not conspicuously unclear.

However, any person may make, to the commissioner of the patent office, a request for a technical opinion as to the registrability of a claimed device in a registered UM or UM application. Where a request is filed, the commissioner will appoint an examiner to prepare a report of a technical opinion as to the registrability of the claim. This is known as a 'Jitsuyoshinan Gijutsu Hyokasho'. This registrability report is equivalent to a substantive examination under the patent law. If the examiner finds out in the course of making the registrability report that the UM registration should never have been registered, the registered right may not be invalidated *ipso facto* by the result of the report. A trial for invalidation or revocation must be demanded in order to cancel the registration. Moreover, the owner of a UM right may not exercise his rights pertaining to his UM against a third party until he gives a warning in the form of a report of a technical opinion.

Although the registrability report is equivalent to a substantive examination under the patent law, it must be distinguished from the hantei judgement under the patent law. A hantei judgement is handed down by a board made up of three appeal examiners and is a report on the scope of the patent right. A registrability report is a report made by an examiner and is a technical opinion as to the registrability of the device. It determines the novelty and inventive steps based on distributed publications, prior art effect of prior application, and the first to file rule. A hantei judgement can be a basis of validation or invalidation of a patent right by the Tokyo High Court since it is a decision in effect by a board. The registrability report can never serve as a basis of a right except when the rights-holder presents this report when warning an alleged infringer. At any time even after the expiry of a UM right, any party may demand a trial for invalidation. The purpose of such proceedings even after the expiry of rights is to make restitution where an alleged infringer has been adjudged guilty based on a right which is subsequently invalidated.

A UM right shall come into force upon registration. The registration and the establishment of right are published in the *Utility Model Gazette*.

Scope of Protection

The Japanese UM law distinguishes between direct and indirect infringement. An object used for the manufacture of the article covered by the registered UM is deemed to be an indirect infringement. The scope of protection extends not only to the object but also to any special moulding or printing machines which have been used to produce the infringing device. Where the machine has other non-infringing uses such as a computer, the owner of the right has the burden of proving that such a machine is used or is to be used for infringement of the device. The owner of a UM right shall have an exclusive right to commercially work the registered UM. Acts of manufacturing, assigning, leasing, importing or offering for assignment or lease of, in the course of trade, articles to be used exclusively for the manufacturer of the article covered by the registered UM shall be deemed to be an infringement.

7.4 THE PERPLEXING DECREASE OF UTILITY MODEL REGISTRATION FIGURES

Does an expansion of the scope of patent rights induce more innovative effort? An analysis was undertaken in 2000 gauging whether the 1987 Japanese patent reforms had had any impact. The reforms had expanded the scope of patents. However, the authors found *no* evidence of a statistically or economically significant increase in either R&D spending or innovative output that could plausibly be attributed to these reforms.[4] Similarly in 1993, due to the fact that there was a heavy delay in the grant of UM applications, the Japanese government revised the UM law to allow accelerated registration without substantive examination, and to shorten the term of protection to six years from the filing date on the basis that product lifecycles were expected to shorten. Table 7.2 shows the rate of applications and grants over a 23-year period.

4 Sakakibara, M., L. Branstetter and M.P. Page (2001), 'Do stronger patents induce more innovation? Evidence from the 1988 Japanese patent law reforms', *Rand Journal of Economics*, accessed at www2.gsb.columbia.edu/faculty/lbranstetter/PDFpapers/Do_stronger_patents_induce_more.pdf.

Table 7.1 Utility model filing in Japan

	Utility model applications	Utility model registrations	Patent applications	Patent registrations
1980	191 785	50 001	191 020	46 106
1981	198 979	50 900	216 307	50 904
1988	171 656	42 300	345 418	55 300
1989	153 277	47 100	357 464	63 301
1990	138 272	43 300	367 590	59 401
1991	114 687	36 500	380 453	36 100
1992	94 601	65 200	384 456	92 100
1993	77 101	53 400	366 486	88 400
1994	17 531	53 885	353 301	82 400
1995	14 886	63 966	369 215	109 100
1996	14 082	95 481	376 615	215 100
1997	12 048	50 108	391 572	147 686
1998	10 917	35 513	401 932	141 448
1999	10 283	21 986	405 655	150 059
2000	9 587	12 613	436 865	125 880
2001	8 806	9 441	439 175	121 742
2002	8 603	7 793	421 044	120 018
2003	8 169	7 694	413 092	122 511

Sources: WIPO Annual Statistics and Annual Report 2000, Japan Patent Office.

Table 7.2 shows the rate of applications and grants of utility models over a 23-year period, as compared to the (almost) inversely proportional rate of patent registrations.

The emerging pattern is a steady drop in applications for registrations from approximately 191 000 (1980) to 77 000 (1993) to 8000 (2003). While the accelerated registration was popular with industry, the legal uncertainty caused by the 'no-examination' rule made the system less satisfactory for business – the technology transfer of unexamined rights not being popular in Japan. It also turned out that many firms were unhappy with the reduced term of protection as there is little chance of succeeding in an action for injunctive relief with a protection period of six years. No doubt, these are key reasons for the failure of the revised system to reverse the fall in applications.[5]

[5] Heath, C. (ed.) (2003), 'Japan', in *Intellectual Property Law in Asia*, London and The Hague: Kluwer Law International.

Another underlying reason for the drop in popularity of the utility model system can be that the Japanese industry has become more innovative and has increasingly been opting for the patent route to legal protection of its inventions. A more cynical view would be that both Japan and the United States have experienced increased patenting due to the expanded scope of patentable subject matter, and due to the relaxed stance of patent examiners (see discussion on the United States in section 3.6).

Individual Versus Company Applicants

According to the Intellectual Property Policy Committee of the Industrial Structure Council, UMs are more popular with SMEs and individuals than are patents. While only 5 percent of patent applications come from individuals, the figure for UMs is 40 percent.

Standard Patents: Japanese Versus Foreign Applicants

Table 7.2 Japanese versus foreign standard patent applications and registrations

	Applications		Registered	
	Residents	Non-residents	Residents	Non-residents
1994	319 938	33 363	72 757	9 643
1995	334 612	34 603	94 804	14 296
1996	340 101	36 514	187 681	27 419
1997	350 807	40 765	129 937	17 749
1998	359 381	42 551	125 704	15 744
1999	360 180	45 475	133 960	16 099
2000	387 364	49 501	112 269	13 611
2001	386 767	52 408	109 375	12 367
2002	369 458	51 586	108 515	11 503

Source: Japan Patent Office.

The overwhelming majority of patent applicants are Japanese residents. However, the proportion of foreign applications has slightly increased from 10.43 percent in 1994, to 11.84 percent in 1998, and 13.96 percent in 2002.

Utility Models: Japanese Versus Foreign Applicants

Table 7.3 Japanese versus foreign utility model applicants

	Applications		Registered	
	Residents	Non-residents	Residents	Non-residents
1994	16 499	1 032	53 086	799
1995	13 685	1 201	62 494	1 472
1996	12 828	1 254	93 409	2 072
1997	10 717	1 331	48 391	1 717
1998	9 669	1 248	34 003	1 510
1999	8 953	1 330	20 595	1 391
2000	7 897	1 690	10 882	1 731
2001	6 988	1 818	7 551	1 890
2002	6 942	1 661	6 216	1 577

Again, the overwhelming majority of applicants are Japanese residents, and the Japanese to foreigner ratio is diminishing more quickly in the case of utility models. The percentage of foreign applications increased from only 6.25 percent in 1994, to 12.91 percent in 1998, and 23.93 percent in 2002. In 2002, out of a total of 8603 applications, 1661 were submitted by foreigners, mostly from Taiwan. As for registrations, out of 7793 in that year, 1577 came from foreign residents.

7.5 COMPARISON WITH THE CONVERSE EXPERIENCE IN KOREA

On a regional level, it is interesting to compare the Japanese utility model culture with South Korea. South Korea has had a UM law since 1961, and the current law is the Utility Model Act 1 July 1999. The law offers UM protection to any technical creation which utilises the rules of nature – this is similar in ethos to the Japanese UM law. The UM must be novel, capable of industrial applicability and possess an inventive step. Novelty is universal.

In parallel with the Japanese development, South Korea withdrew substantive examination from UMs in 1999. Anyone can, however, file a request for the 'technical evaluation' of a UM application or registration i.e. a detailed examination report. An evaluation report, at any event, is a prerequisite to any action to enforce UM protection. The right to exercise the UM right can then only proceed if the registered utility model owner sends a warning letter to the offender including a certified copy of the 'technical evaluation'.

If we look at the application and grant rate under the South Korean UM regime, we find however, a very healthy set of statistics. Table 7.4 shows that for the last three years, Korea has experienced a steady rate of approximately 35 000 applications per year, as compared to the Japanese figure of approximately 8500 per year.

Table 7.4 Utility model statistics for South Korea from 1992 to 2002

	Applications	Grants
1992	28 665	7 870
1993	32 205	7 592
1994	39 790	7 817
1995	59 856	8 148
1996	68 822	9 191
1997	45 809	13 713
1998	28 890	25 715
1999	30 650	32 868
2000	37 163	41 745
2001	40 804	43 842
2002	39 187	39 955

The Korean Intellectual Property Office statistics note that foreign applications are negligible compared to domestic ones. From 1998 to 2002, out of the 176 700 applications submitted to the Korean Intellectual Property Office, only 1889 came from abroad, which is little more than 1 percent of the total. Registration figures tell a similar story, with only 2332 foreign ones out of the 184 127 registered during the same time period. The Korean experience is interesting in that its domestic sector is active both in utility model and patent filing. Of course, compared to utility models, foreign patenting activity takes place at a much higher rate. However, overall, domestic patent filing and registration rates are far higher.[6]

[6] Out of the total of 467 892 patent applications filed between 1998 and 2002, 138 222, or just under 30 percent, came from foreigners. The percentage of foreign patents granted was only slightly higher, with 76 298, or 33 percent of the total of 230 456.

Table 7.5 Utility model statistics for South Korea from 1992 to 2002

	Applications	Grants
1992	28 665	7 870
1993	32 205	7 592
1994	39 790	7 817
1995	59 856	8 148
1996	68 822	9 191
1997	45 809	13 713
1998	28 890	25 715
1999	30 650	32 868
2000	37 163	41 745
2001	40 804	43 842
2002	39 187	39 955

7.6 CORRELATION BETWEEN UTILITY MODEL AND INNOVATION CULTURE?

It may be that there has been a shift in Japanese filing behaviour in the past ten years due to the shift in the innovation culture in Japan. Japanese industries tended to focus on incremental innovation rather than radical innovation during the period from the post-war years to the 1980s,[7] and this trend has reversed. This, in turn, meant that the UM system was not seen to be as vital as it had once been.[8] The important lessons from Japan are that considering that Japan was the first Asian country to introduce the UM regime into its domestic law, this feature may have enhanced its industrial and economic capability in the early twentieth century. However, innovation may not necessarily have been encouraged by non-examination – a feature which was only introduced in 1993, when UM registrations were on the decline. The drop in registration figures may be a consequence of the change in her industrial and innovation policy.

In South Korea, the central government's innovation policy has been focused on SMEs and it has recently resulted in a 'regional innovation support system'. Moreover, South Korea's industrial, technology and regional policies have changed so that her economy has been re-structured from low-technology,

[7] Mansfield, E. 'Industrial R&D in Japan and the United States: A Comparative Study', 1988, p. 326.

[8] Heath, C. (2002), 'Utility model law', in *Encyclopaedia of Japanese Law from 1868*, Netherlands: Brill Publishing, (copy on file with author).

labour-intensive, 'mass production' types of industry to high-technology, capital- and skill-intensive, 'flexible specialisation' types of industry. Conversely, it is also reported that one of the weaknesses in South Korea's innovation system is the lack of technological spin-offs and the lack of transfer of basic research results technology to transfer research results from public research establishments to industry and particularly to SMEs.

On a more cautious note, however, we should note that filing statistics alone do not explain macroeconomic trends, and further research is needed to explain why the figures are falling in Japan but not in Korea despite similar laws and reforms.

8. China and Taiwan

Kit Boey Chow, Kah Mun Leo, Susanna Leong with Jerry Hsiao

Taiwan and the People's Republic of China have both adopted UMs[1] in their patent legislation. Despite being ethnically similar, the two countries have a divergent past in dealing with intellectual property protection. Taiwan industrialised earlier than China and faced successive trade pressures from the United States. Due to this, Taiwan developed her patent law in compliance with the international norm. China, on the other hand, followed more socialist principles and held that the grant of private monopoly rights to an individual or a firm was contradictory to socialist principles; in fact, the government has openly endorsed imitation as the way to develop the economy of the country. Nowadays, however, both Taiwan and China as members of the WTO, have embraced a more international standard of TRIPS, and even, in the case of UMs, both countries have adopted a TRIPS-plus standard.

8.1 CHINA

8.1.1 Economic and Innovation Climate

China has experienced a paradigm shift and has now become a common outsourcing manufacturing centre for the United States, Japan and Europe. Tremendous growth has been witnessed in recent years after establishing its open door policy towards foreign trade and investments, and improving its economic and legal structures. Advancements in the area of information and telecommunications have made China a competitive nation in the post-WTO accession era based on its abundance of cheap labour and natural resources.

Despite the remarkable economic growth and development, innovation seems to be rather limited as most of the private sector do not possess the necessary

[1] The terms utility model patent, petty patent, registered inventions or second tier protection are used interchangeably.

financial and human resources to conduct their own R&D activities.[2] It is of interest to note that the industrial infrastructure of China, until 1978, was dominated by large state-owned enterprises (SOEs). Since then much of the boom in manufacturing output has been produced by 'collective' enterprises under the aegis of local governments – particularly the township and village enterprises (TVEs) or, increasingly, by private entrepreneurs or foreign investors, either in wholly owned enterprises or in joint ventures with Chinese interests. By 2002, the share of state-owned and state-holding enterprises in gross industrial output value had shrunk to 41 percent and represented only 16 percent of industrial output.[3]

The strengths that are visible within the Chinese economy are its industrial restructuring and strengthening of economic competitiveness. Weaknesses included a critical need for improvement in relevant laws and regulations, and weakness in some industries with respect to international competitiveness. One World Bank report stated, amongst its many recommendations, that China should institute an economic regime that offers incentives for the efficient use of existing knowledge and creation of knowledge and entrepreneurship.[4]

Throughout the drafting of the Patents Act, strong debates were held regarding the adoption of the UM law (*shiyong xinxing*), with legislators fearing that the patent office would be flooded by minor inventions from foreign corporations, especially those from Japan. However, in the end, the UM patent was enacted and the results have shown that it has been wildly successful with local innovators, with the highest number of applications coming from Chinese-based individuals and firms.

8.1.2 China's Three-Tier Patent System

Historical background

The adoption of Western style intellectual property laws in China originated in 1898 during the Qing dynasty. The law was aptly called the 'Regulations on Rewards for the Promotion of Technology', though this was soon replaced, in 1911, by the 'Interim Charter on Rewarding Industrial and Artistic Producers'. After the downfall of Imperial China, the first Patent Act was enacted in 1944 by the Kuomintang government; this law, although forming the basis of future

---

[2] Cong C. (2005) *Corporate R&D and Innovation in China*, East Asia Institute, National University of Singapore.

[3] *The Economist* (2004), 'China's economic structure', 9 February, accessed 15 November, 2005 at www.economist.com/countries/China/profile.cfm?folder=Profile-Economic%20Structure.

[4] Dahlman, C. and J-E. Aubert (2000), *China's Development Strategy, The Knowledge and Innovation Perspective*, World Bank Institute.

Taiwanese patent laws (see page 163), never came into force in mainland China (later the People's Republic of China in 1949).[5] The modern form of intellectual property protection came into force in China during the 1950s, under the title of 'Provisional Regulations on the Protection of the Invention Rights and Patent Rights'. However, only four patent certificates and six invention certificates were issued between 1953 and 1957, and this system was abolished during the Cultural Revolution.[6]

China's first modern patent law was enacted in 1984 and came into effect in 1985. Since then, the law has been amended twice. The first revision, undertaken in 1992, extended the patent length from 15 to 20 years for invention patents and from five to ten years for patents on utility models and external designs. The second revision, which was completed in September 2000, eliminated the provisions under the old law that prevented state-owned enterprises from trading their patents in technology markets. The second revision also introduced new provisions designed to make it more rewarding for enterprise employees to innovate. Since the passage of the 1984 patent law, the central government has issued over 20 regulations and guidelines to promote innovative activity in China. Today's patent law in China is pretty much in line with the international standard. Up until now, China has acceded to all the international patent treaties, and its laws on intellectual property rights meet the requirements of the WTO's Agreement on Trade-related Aspects of Intellectual Property Rights (TRIPS).

Substantive law

Three types of patent protection are available in China, namely the invention patents, utility models and design patents respectively:

- *invention (standard) patents* which are conventional patents, with a protection period of 20 years from the date of filing or the priority date. Excluded from invention patents are scientific discoveries, rules and methods for diagnosis or the treatment of diseases, animal and plant varieties and substances obtained by nuclear transformation;
- *design patents*, which are conferred on original shapes, patterns, colours or a combination of these elements in an object. They have a life-span of ten years from the date of application or priority date;
- *utility models*.

---

[5] Ganea, P. (2005), 'Patents, utility models and designs', in P. Ganea, T. Pattloch and C. Heath (eds), *Intellectual Property Law in China*, The Hague: Kluwer Law International, pp. 1–2.

[6] Fai, F.M., 'China's growing technological capabilities: opportunities and threats for foreign multinational enterprises', accessed 9 November 2005 at www.iacmr.org/04-022.pdf.

Under China's Implementing Regulations, 'utility model' in the patent law means any new technical solution relating to the shape, the structure, or their combination, of a product, which is fit for practical use. A UM patent is therefore only granted for product-related technical solutions but not for processes, biological material, fluids, gases, or computer-implemented invention-creations.[7] The UM patent is subject to the following patentability requirements: novelty, inventive step and industrial applicability.

Novelty means that before the filing date no identical invention or utility model has been publicly disclosed, used or made known to the public anywhere in the world.[8] Despite this, as we will see, UMs are being granted to local 'inventors' for inventions imported from overseas. Allegedly, this is giving rise to perverse behaviour by some UM owners – if true, a good justification for universal novelty to be applied strictly.

Section 22 (3) of the Patents Act defines inventive step by stating that 'as compared with the technology existing before the date of filing, the invention has prominent substantive features and represents a notable progress and that the UM has substantive features and represents progress'. It is currently confusing to distinguish the difference between 'prominent substantive features' and 'notable progress' and 'substantive features' and 'progress'. According to the Higher People's Court of the Beijing Municipality, 'prominent substantive features' means an invention with essential technological breakthrough, resulting in prominent and essential changes in the relevant field of technology; 'notable progress' represents a great leap forward which overcomes any shortcoming of the relevant technology, or a noticeable technical success.[9] Thus, in the Examination Guidelines, the following solutions were proposed: in the case of a standard patent, an invention is deemed to be non-obvious even to an expert who has conducted a comprehensive search in all the neighbouring and related fields but in the case of a UM, the search should be restricted to the field to which the technical solution immediately pertains or only a few neighbouring fields.[10]

---

[7] Fai, F.M. 'China's growing technological capabilities: opportunities and threats for foreign multinational enterprises', p. 9.

[8] According to Sec. 22 (2) novelty is absolute novelty that, 'before the date of filing, no identical … utility model has been publicly disclosed in publications in the country or abroad or has been publicly used or made known to the public by any other means in the country, nor has any other person filed previously with the Patent Administration Department under the State Council an application which described the identical invention or utility model and was published after the said date of filing'.

[9] Fai, F.M. 'China's growing technological capabilities: opportunities and threats for foreign multinational enterprises', p. 17.

[10] Id. p. 18.

Unlike Taiwan, Sec. 22(4) of the Chinese Patent Act states that a UM should deem to be industrially practicable i.e. the invention or UM can be made or used and can produce effective results. This requirement is broader than the industrial applicability requirement required under the TRIPS and is a deliberate attempt by the government to develop its economy and to enable more inventions and UMs to qualify for a patent. The following are deemed not to be practically applicable:

- The invention cannot be repeatedly realised.
- The realisation of the invention is confined to certain unique natural environments.
- The invention cannot be explained by natural law.
- The subject matter of the invention is a non-technical method other than medical treatment.

UM applications are examined *ex officio* as to compliance with formalities only, and if there is no cause for objection of the application on formalities, the patent office issues the relevant patent certificate and publishes it. However, in litigation relating to UMs, the tribunal before which the litigation is pending may require the owner of the UM to submit a search report issued by the patent office relating to the novelty of the subject matter of the UM. The processing time for a grant under the UM system typically ranges from six months to a year, as against one to four years for an invention patent. A UM is more attractive as it costs around 40 percent[11] of the invention patent where technologies are rapidly changing.

8.1.3 Empirical Analysis of the Utility Model System

Out of a total of 1 569 324 patent applications made in China between 1994 and 2003, about two-fifths (41.3 percent) of them were UMs, of which almost all were local applications. In contrast, foreign applications constituted the majority for standard patents. Applications for UMs started to increase faster in 2000, reaching 109 115 cases in 2003. The composition of applications by type and ownership is as shown in Table 8.1.

[11] *http://www.chinasinda.com/ncat.2.php?id=132.*

Table 8.1 Composition of applications by type and ownership

Type of application	Total applications	Local applicants	Foreign applicants
Utility model	648 665	644 038 (99.3 percent)	4 627 (0.7 percent)
Standard patents	476 041	226 674 (47.6 percent)	249 367 (52.4 percent)
Design patents	444 618	407 338 (91.6 percent)	37 280 (8.4 percent)

Accumulated patent awards totalled 885 399 or about 56.4 percent of total applications (Figure 8.2). About half (50.1 percent) of the awards were UMs. The breakdown was:

- 443 561 utility models (50.1 percent); local 440 571 (99.3 percent), foreign 2990 (0.7 percent)
- 113 722 invention patents (12.8 percent); local 39 700 (34.9 percent), foreign 74 022 (65.1 percent)
- 328 116 design (37.1 percent); local 296 552 (90.4 percent), foreign 31 564 (9.6 percent).

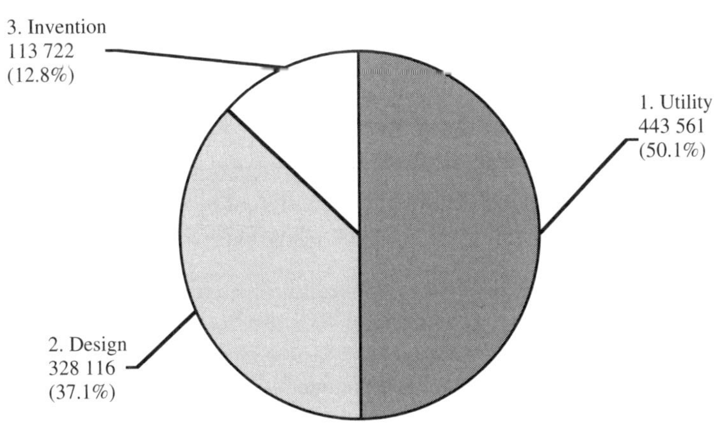

Source: State Intellectual Property Office (SIPO), Annual Report 2003.

Figure 8.1 China accumulated patent awards by categories, 1994–2003

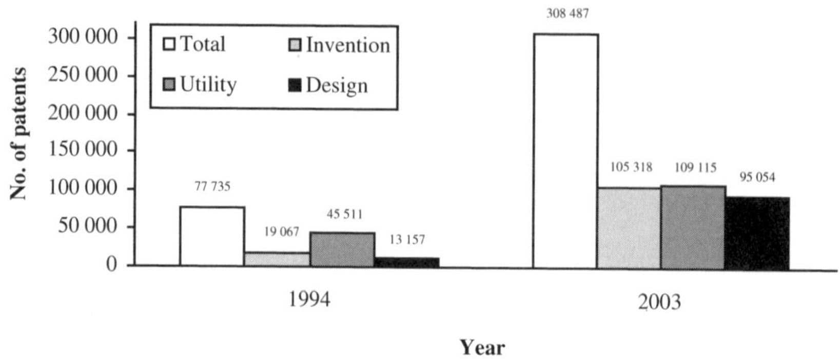

Source: State Intellectual Property Office (SIPO), Annual Report 2003.

Figure 8.2 Patent applications by type in China, 1994 vs 2003

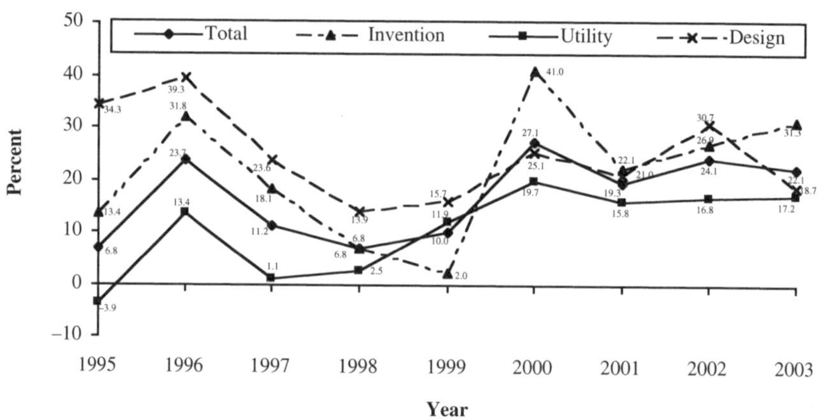

Source: State Intellectual Property Office (SIPO), Annual Report 2003.

Figure 8.3 Growth in patent applications in China, 1994–2003

Between 1994 and 2003, total patent applications rose from 77 735 to 308 487 or at an annual compounded growth rate of 16.5 percent (Figure 8.2). In the nine year period, design patents witnessed the fastest growth from 13 157 to 95 054 applications, representing an annual compounded growth of 24.6 percent. They also had the highest growth per annum except for the years 2000 and 2003[12]

[12] The highest annual growth in applications for the years 2000 and 2003 was achieved by invention patents.

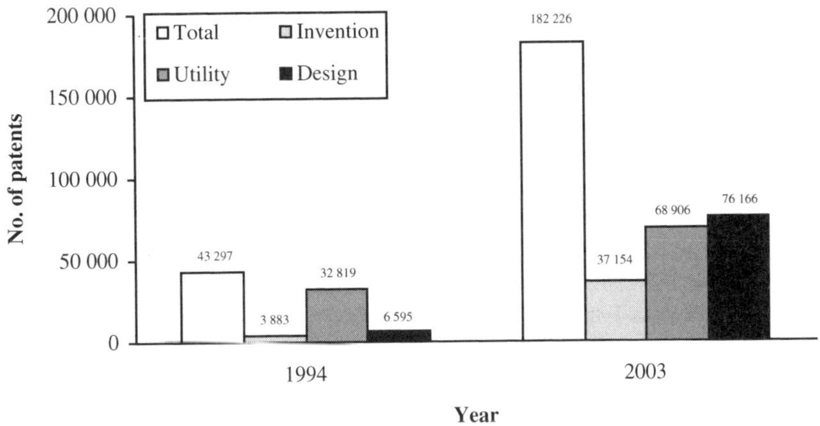

Source: State Intellectual Property Office (SIPO), Annual Report 2003.

Figure 8.4 Patent awards by type in China, 1994 vs 2003

(Figure 8.3). The annual compounded growth for invention patents was 20.9 percent and for utility models 10.2 percent.

Total patent awards grew from 43 297 to 182 226 or at a compounded rate of 17.3 percent a year (Figure 8.3). Fastest growth was witnessed by design patents from 6595 to 76 166 awards or at 31.2 percent per annum. Other patent award growth rates were 28.5 percent for invention patents and 8.6 percent for utility models.

The surge of statistics can be explained partly by the fact that counterfeiters are now employing the UM and design patent systems to claim protection for their modified versions of goods and products which are protected only under foreign patents.[13]

Applications for UMs started to increase faster in 2000, reaching 109 115 cases in 2003 (Figure 8.5). The pattern for UM awards differed, falling in 1995–1997, reaching a peak in 1999 and hovering around 55 000 a year in the following three years. The year 1995 marked a decline in both UM applications and awards. The ratio of awards to applications was the highest in 1999 at 98 percent, as against 72 percent in 1994 and 63 percent in 2003.

Compared to locals, foreigners achieved higher growth in both UM applications and awards between 1994 and 2003 (Figures 8.7 and 8.8):

[13] Spurgeon, Brad (2004), 'The new Chinese counterfeit game', *International Herald Tribune*, 15 November.

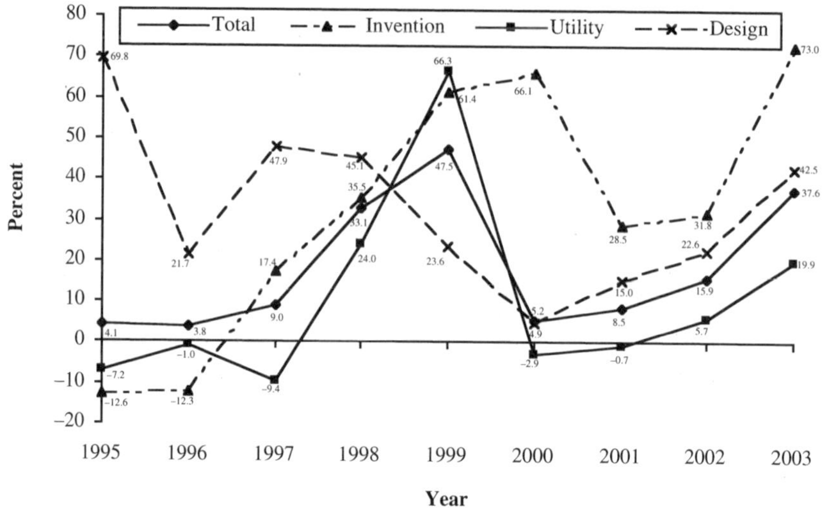

Source: State Intellectual Property Office (SIPO), Annual Report 2003.

Figure 8.5 Growth in patent awards in China, 1994–2003

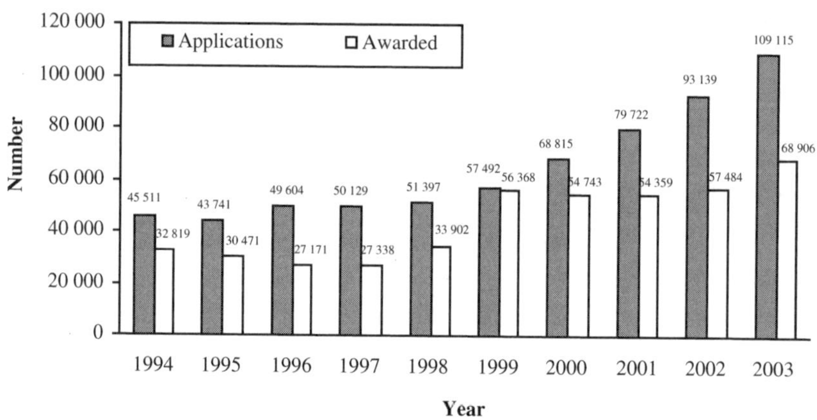

Source: State Intellectual Property Office (SIPO), Annual Report 2003.

Figure 8.6 Utility model applications and awards in China, 1994–2003

- applications – foreign 16.5 percent per annum, local 10.2 percent per annum
- awards – foreign 12.8 percent per annum, local 8.6 percent per annum.

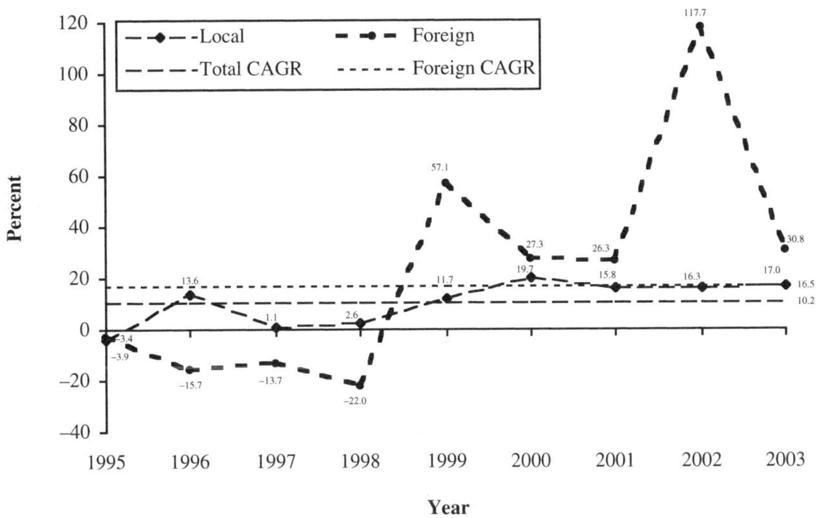

Source: State Intellectual Property Office (SIPO), Annual Report 2003.

Figure 8.7 Utility model applications growth by ownership in China, 1994–2003

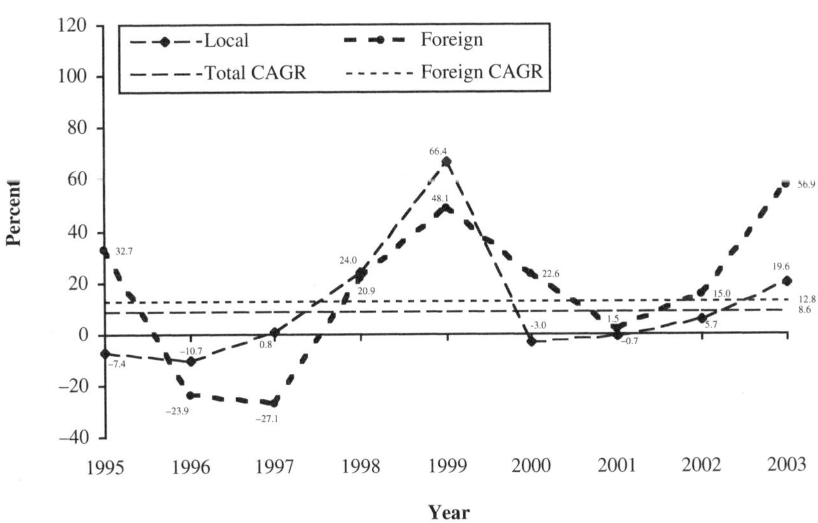

Source: State Intellectual Property Office (SIPO), Annual Report 2003.

Figure 8.8 Utility model awards growth by ownership in China, 1994–2003

The higher growth in foreign ownership of UMs was due in part to a small base of less than 1 percent in absolute numbers for both applications and awards respectively for each year between 1994 and 2003. Therefore, growth in overall UMs was basically determined by the domestic market.[14] Apparently such second tier innovation was put to good use by smaller local enterprises, implying a lower level of high value innovation by domestic inventors.

8.1.4 Policy Implications

No economic or statistical analysis of UMs in China would be complete without mentioning that many UMs are also being granted to local 'inventors' for inventions imported from overseas. There are two alternative possible explanations for the statistics.

One explanation can be that the surge in UM applications reflects the growth in the innovative sectors in China, and a growing awareness of intellectual property protection and rights. Apparently such second tier innovation has been put to good use by smaller local enterprises, implying a lower level of high value innovation by domestic inventors. An alternative explanation is that these statistics reflect the fact that counterfeiters are now employing the UM and design patent systems to claim protection for their modified versions of goods and products which are protected only under foreign patents. Allegedly, this is giving rise to perverse behaviour by some UM owners who are threatening to take legal action against the true inventors who are often foreign companies seeking to expand their commercial activities in China. If true, a strong case can be made to argue that universal novelty should be applied strictly not only by the granting office but also by the courts.[15]

Finally, as one commentator indicated, although the UM system is viewed as being valuable in China, problems exist within this system. In his view, the fact that the Chinese system offers UM protection without substantive examination means that it is difficult to guarantee the quality or level of inventiveness involved in a UM. Consequently disputes have occurred. Thus, the author suggests that this can be resolved by insisting that all applications for UMs provide a corresponding search report.[16] Nevertheless, this will increase the costs of the

---

[14] This is the reason why Figures 8.7 and 8.8 cover only the local utility patents growth rates since they constitute more than 99 percent of the total utility patents every year from 1994 to 2003.

[15] Spurgeon, B. (2004), 'The new Chinese counterfeit game', *International Herald Tribune*, 14 November.

[16] Shoukang, G. (1997), 'The development and perspective of intellectual property in the People's Republic of China', *Intellectual Property Quarterly*; also Heath, C. (1998), 'Utility models in East and West', in *Current Problems of Intellectual Property*

system. Moreover, as the number of patent applications and patents granted increases quickly, the requests for re-examination and invalidation also increase. A Re-examination Board was set up in China to deal with appeals filed by applicants who were not satisfied with the decision of the lower examining board. The board also deals with patent invalidation requests, and the rate of invalidation requests has been high, making up more than 40 percent of the total. One Chinese Patent Office examiner opined that the reason for the high invalidation rate is because most invalidation proceedings relate to UMs i.e. 95 percent of the invalidation requests were filed against patent rights for UMs, and more than 60 percent of the cases ended in the invalidation of the UM right.[17] Yet another Chinese attorney laments that the UM system is being abused since no examination as to substance is made, either at grant stage or even at infringement stage (it remains an option for the court or administrative authority to ask the patentee to furnish a search report).[18]

8.2 TAIWAN

8.2.1 Historical Background

The first Patent Act, introduced into Taiwan on 1 January 1949, offered invention patents, UM patents and design patents. The Act has been revised several times, with the latest amendments in 2003,[19] which introduced several substantial amendments to the UM patent. The main purpose of the UM patent is to protect smaller inventions, and the system adopts an inventive step requirement which is lower than that of an invention patent.[20]

UM patent is popular in Taiwan because 98 percent of Taiwanese businesses are small and medium enterprises with very diverse R&D capability[21] who

Law – Writings in Honour of Nobuo Monya, Tokyo, pp. 47–72, discussing the salient aspect of the Chinese utility model system.

[17] Yin Xin-Tian (1991), 'A patent invalidation case in China?', 21 IIC 513, p. 514; Li Zhen (1998), 'Patent appeal system in China', presentation to the International Patent Appeal Examination Symposium, organised by the Japanese Patent Office, 19–20 November, Tokyo, accessed at www.jpo.go.jp/shiryou_e/toushin_e/kenkyukai_e/21stcen.htm.

[18] Ye, Mark (2004), 'Patents for utility model and design be renamed patent tegistration', 4 China Patents and Trademarks 34.

[19] Hsu, H. (2003), 'Patents', in C. Heath (ed.), *Intellectual Property Law in Taiwan*, Max Plank Series on Asian Intellectual Property Law, Hague: Kluwer Int., pp. 22–4.

[20] The Court supports this lower-level inventive step approach. See Administrative Supreme Court Decisions: No. 83 Pan Tzu 2875, 30 December 1994 and 79 Pan Tzu 2087, 28 December 1990.

[21] 'Discussing the importance of establishing a research and development alliance for Taiwan's industries', interview with Director Huang Chong-Qiu, Department of In-

favour a registration system which grants protection rapidly. For example, Taiwan has very mature semiconductor, LCD and computer technology industries that have relatively short product life-spans, and such products require a registration system which grants large numbers of manufacturing process and product patents in a short time.[22] Indeed, Taiwanese nationals are heavy users of UM patents. In the year 2002, for example, amongst the 24 220 patents filed by Taiwanese applicants, only 9170 were for invention patents, 6820 were designated for design patents whereas 24 200 (60 percent of all applications) were for UM patents.

8.2.2 Substantive Law

Similar to the Japanese Patent Law, Taiwan's Patent Law does not allow applicants to make two applications in relation to the same invention. An applicant must choose whether to pursue an invention patent or a UM patent. Although many aspects of patent law requirements apply to UM law *mutatis mutandis*, there are some fundamental differences between invention patent and UM patent in Taiwan.

First, patent protection covers a method, substance or device whereas a UM is defined as a creation or an improvement that has been made in respect of a form, construction or shape of an object.[23] This definition includes both two- and three-dimensional structures, appearances of an object which may be defined by lines and planes and which is functional, the internal or integral structure of an object expressed either by arrangements or layouts or the relation between components within that object.[24] Excluded from UM protection are inventions which are contrary to public order, good custom or hygiene; shapes which are identical with or similar to that of the flag of a political party, national flag, military flag, a national emblem or a government medal, and methods or processes.[25]

A second difference between invention patents and UMs is that the former are granted 20 years of protection from the date of filing whereas UMs are

---

dustrial Technology, Ministry of Economics. Available online: www.stpioneer.org.tw/modules.php?name=magazine&pa=showpage&tid=2250 (on file with the author).

22 Kung, M.H. and H.Y. Lin (2003), 'The dynamic of innovation in Taiwan – the patent analysis respective', conference paper presented to the International Conference on 2003 industrial science and technology innovation-policy issues and challenges in a critical era, 30–31 October, Taipei, Taiwan.

23 Taiwan Patent Act, Sec. 93.

24 Intellectual Property Office: Guidelines for Patent Examination.

25 Taiwan Patent Act, Sec. 96 (this part 'a shape identical with or similar to that of the flag of a political party, national flag, military flag, a national emblem or a government medal' is abolished).

granted ten years of protection. A third difference between the two laws is that the criteria for protection under UM law are novelty, inventive step and industrial applicability. Taiwan adopts an absolute novelty principle.[26] However, the inventive step is lower for a UM application than that of an invention patent; inventiveness is present for the purposes of an invention patent if the invention could not have been easily made by those skilled in the art, whereas inventiveness is present in a UM unless it is obviously easy for those skilled in the art to make the device.

Compulsory licensing is available for patents but not for UMs. A final difference between the two laws is that substantive examination was no longer required, after 7 June 2004,[27] for UM applications.

The major 2003 revisions which affect UM law are as follows:

(a) *Abolition of substantive examination*: UMs are no longer to be examined for novelty and inventive step prior to grant. This amendment was made to meet the demands from the industry for a faster protection regime for products with short lifecycles. Specifically, the industry was of the view that UM patents would fail in promoting innovation from SMEs if they were to be subjected to substantive examinations. The new revised UM law, which relies heavily on the Japanese Utility Model Law, attempts to strike a balance between the interests of the patentee and the legal certainty of transaction by making substantive examinations available during any invalidation procedures.

(b) *Introduction of the 'Technical Evaluation Report'*: This new feature allows anyone who is uncertain as to the enforceability of a UM patent to apply for a technical evaluation report from the competent authority such as the Taiwan Intellectual Property Office (TIPO) on a claim-by-claim basis. The purpose of the evaluation report is to act as an early warning signal when exercising the patentee's rights, in order to avoid abuse of the UM system and to stimulate public inspection. The usual procedure is for TIPO to grant the UM right without a substantive examination but merely a verification of formalities. Should the rights holder wish to enforce his UM, he should request a technology report as soon as possible. For the technology report, TIPO appoints an examiner to evaluate the novelty and inventiveness of the product. The technical evaluation report does not have any legally binding effect nor is it an administrative order. The report is merely a reference for exercising the UM patent rights or a refer-

26 Taiwan Patent Act, Sec.95.

27 Hsu, H. (2003), 'Chapter 3: Utility models' in C. Heath (ed.), *Intellectual Property Law in Taiwan*, Max Plank Series on Asian Intellectual Property Law, Hague: Kluwer Law Int., p. 119.

ence for the patentee on 8 April 2005, the first technical evaluation report in Taiwan was issued.

8.2.3 Empirical and Policy Analysis

Table 8.2 Statistics of utility model patents in Taiwan (2001–2003)

Year	Application	Re-examination	Rejection	Approval	Certificate issued
2001	25 370	2 797	9 668	21 212	16 680
2002	21 750	3 100	9 453	16 115	15 200
2003	21 935	3 551	11 165	21 439	15 505

Source: Annual Statistics Taiwan Intellectual Property Office.

Table 8.3 Statistics on domestic and foreign patent application in Taiwan (2001–2003)

Year	Domestic	Foreign
2001	24 220	1 150
2002	20 962	1 058
2003	21 231	704

Source: Annual Statistics Taiwan Intellectual Property Office.

Table 8.4 Statistics on domestic and foreign patent approval in Taiwan (2001–2003)

Year	Domestic	Foreign
2001	19 999	1 213
2002	15 265	850
2003	20 135	1 124

Source: Annual Statistics Taiwan Intellectual Property Office.

In the year 2003, the top five foreign countries to receive UM protection in Taiwan were:

1. Japan (508)
2. United States (299)
3. Germany (24)
4. South Korea (32)
5. Netherlands (24).

In year 2003, the top five domestic industries to make applications for UM protection were:

1. transporting (2344)
2. basic electronic elements (2258)
3. working of metal (1962)
4. domestic articles (1933)
5. pharmaceutical and entertainment (1664).

In year 2003, the top five foreign industries to make applications for UM protection were:

1. basic electronic elements (188)
2. transporting (157)
3. pharmaceutical and entertainment (148)
4. domestic articles (117)
5. optics (56).

We can see from the above tables and statistics that Taiwanese nationals use UMs rather heavily as compared to the foreign applicants. However, does the high issuance of UMs create more disputes? Judging from Table 8.5 which shows the statistics in relation to cases heard and disposed of in the national patent office, this was not the case. The numbers in Table 8.5 indicate the total number of applications made in relation to invention patents, UMs and design patents. Despite the high numbers of applications, the table indicates a relatively low invalidation rate – this may change as these figures reflect the fact that prior to June 2004, the Intellectual Property Office examined all UM and invention patents substantially.

Table 8.5 Statistics on cases filed and disposed of patent (2001–2003)

Year	Application	Re-examination	Opposition	Invalidation
2001	67 860 (25 370 related to utility models)	10 463	2 596	701
2002	61 420 (of which, 21 750 related to utility models)	11 728	1 734	591
2003	65 742 (of which 21 935 related to utility models)	13 325	1 876	512

Note: Figures reflect the number of cases applied in that year.

Source: Annual Statistics Taiwan Intellectual Property Office.

9. The ASEAN states

Uma Suthersanen with Lim Heng Gee

The ASEAN countries have made significant efforts to overhaul their intellectual property systems in the last ten years. At the ASEAN summit in Bangkok in December 1995, an Intellectual Property Framework Agreement was concluded. In 1998, the ASEAN Framework Agreement on Intellectual Property Cooperation was signed. One of the objectives of the ASEAN Framework Agreement is the possibility of setting up an ASEAN patent system and Patent Office to promote the region-wide protection of patents.[1] Harmonisation of rights within the ASEAN region was further imposed by the international obligations of all the member countries, especially in connection with the TRIPS Agreement. Hence, there is some similarity of intellectual property laws within this region.

Nevertheless, as is the situation all over the world, the area of second tier patent systems is as yet unharmonised within the ASEAN region. All ASEAN countries have implemented the UM system except Brunei, Myanmar, Laos and Singapore.

9.1 MALAYSIA (LIM HENG GEE)

On the formation of Malaysia in 1963, there were three pieces of patent legislation in existence, the Registration of United Kingdom Patents Act, 1951 for the Federation of Malaya, the Patents Ordinance, Chapter 61, for the State of Sarawak and the Registration of United Kingdom Patents Ordinance, 1937, for the State of Sabah. All these three were based on the simple re-registration system under which an applicant desiring patent protection would have to make an application to the United Kingdom Patent Registration Office. On being granted a United Kingdom patent, the owner would have to re-register his patent with the Registrar of Patents in the component states of Malaysia (i.e. Federation of Malaya, Sabah and Sarawak) under the different ordinances before he was entitled to patent protection throughout the whole territory of Malaysia. Obviously

1 Article 1.4.

this situation was undesirable, leading to unnecessary expense and duplication of efforts. Thus, on 28 November 1983, the Patents Act 1983 was enacted.[2] By section 89 of this Act, all the previous patent legislation just mentioned was repealed. However, it was not implemented until 1986, after it was further amended by the Patents (Amendment) Act 1986. Further amendments were brought about in 1990, 1993, 2000, 2002 and 2003.

Under the present Patents Act 1983 (as amended), two types of protection are available – the first is through the grant of a patent, and the second is through the issue of a certificate for a utility innovation.

The normal patent protects inventions which satisfy the normal requirements of novelty, inventive step and industrial application. The second system of protection available under the Act aims to protect 'minor inventions', called 'utility innovations' in the statute, whereby a lower level of patentability criteria needs to be satisfied. The provisions governing the grant of protection for utility innovations are laid down in Part IVA of the Act. The Act provides that with the exception of certain specific exclusions, the provisions of the Act, subject to the modifications in the Second Schedule of the Act, are applicable to utility innovations in the same manner as they apply to inventions.[3]

9.1.1 Substantive Features of the Utility Model System

The following is a summary of the Malaysian utility innovation regime:

- no requirement for inventive step
- the application can only contain one claim
- the duration of protection is for 20 years
- need to show that the invention is in commercial or industrial use in Malaysia
- not subject to compulsory licence
- lower registration and maintenance costs.

Under the Malaysian Patents Act 1983 (as amended), two types of protection are available – the first is through the grant of a patent, and the second is through the issue of a certificate for a utility innovation. The latter system of protection available under the Act aims to protect 'minor inventions', called 'utility innova-

---

[2] Act 291, which came into force on 1 October 1986.

[3] Section 17A(1) of the Act (as amended in 1986 and 1993). S. 17A(2) of the Act provides specifically that Sections 11 (definition of a patentable invention), 15 (definition of inventive step), 26 (requirement for unity of invention), Part X (provisions for compulsory licences) and Sections 89 (repeal and saving provisions) and 90 (transitional provisions) shall not apply to utility innovations.

tions' in the statute, whereby a lower level of patentability criteria needs to be satisfied. A utility innovation is defined in the Act as:

> any innovation which creates a new product or process, or any new improvement of a known product or process, which is capable of industrial application and includes an invention.[4]

A certificate for a utility innovation can be granted if the utility innovation is new and is capable of industrial application. There is no requirement for an inventive step, this being specifically excluded by the Act.[5]

The procedure involved in an application for a certificate for a utility innovation, except in relation to the number of claims permissible, is the same as that for a patent for an invention. Unlike a normal patent, where more than one claim can be applied for, in the case of a utility innovation only one claim is allowed. Two forms of substantive examination are available, a full substantive examination or a modified substantive examination,[6] although it is not possible for an applicant to be granted both a patent and a certificate for utility innovation for the same invention.[7] However, it is possible to convert an application for a patent into an application for a utility innovation and vice versa.[8]

A certificate for a utility innovation expires ten years from the filing date.[9] However, before the expiration of this ten-year period, an application for extension for two additional five year periods of protection can be made. This means that the total possible length of protection is 20 years from the filing date of the application, which is the same period available for a normal patent. However, before such extensions can be granted the owner has to file an affidavit showing that the utility innovation is in commercial or industrial use in Malaysia. Failing this, he has to give satisfactory explanations for its non-use.[10]

The limitation period to bring proceedings in the case of patent infringement is five years from the act of infringement. However, in the case of proceedings for infringement of a utility innovation, the period is shortened to two years.[11]

---

4 Section 17 of the Malaysian Patent Act (as amended).

5 Section 17A(2) of the Malaysian Patent Act. However, Section 14(3) provides a 'period of grace' enabling an application to be filed after the utility innovation has been disclosed to the public as a result of acts by the applicant or his predecessor in title, without such disclosure being considered prior art against the application.

6 Section 29A, Malaysian Patent Act.

7 Section 17C, which was added to the Malaysian Patent Act by Section 9 of the Patents (Amendment) Act 1993.

8 Section 17B(1) and (2), Malaysian Patent Act.

9 Section 35, as modified by the Second Schedule, of the Act.

10 Section 35(1) as modified by the Second Schedule.

11 See Section 59(3) as modified by the Second Schedule.

Further, in relation to judicial assignment, where it is alleged that the essential elements of an invention claimed in a patent application or patent have been unlawfully derived from an invention for which the right to the patent belongs to another person, the limitation period to bring an action is up to five years from the date of the grant. In the case of a utility innovation, the limitation period is only three years.[12]

Table 9.1 Comparative costs of application

	Patent	Utility innovation
Application fees	RM 200	RM 100
Request for substantive examination	RM 700	RM 700
Certificate of grant	RM 150	RM 100
Annual fees for 2nd yr after grant	RM 200	–
Annual fees for 3rd yr after grant	RM 250	RM 120
Annual fees for 4th yr after grant	RM 300	RM 160
Annual fees for 5th yr after grant	RM 350	RM 160
Annual fees for 6th yr after grant	RM 400	RM 200
Annual fees for 7th yr after grant	RM 450	RM 200
Annual fees for 8th yr after grant	RM 500	RM 240
Annual fees for 9th yr after grant	RM 550	RM 240
Annual fees for 10th yr after grant	RM 600	RM 280
Annual fees for 11th yr after grant	RM 650	RM 400
Annual fees for 12th yr after grant	RM 700	RM 600
Annual fees for 13th yr after grant	RM 800	RM 700
Annual fees for 14th yr after grant	RM 900	RM 800
Annual fees for 15th yr after grant	RM 1 000	RM 1 000
Annual fees for 16th yr after grant	RM 1 200	RM 1 100
Annual fees for 17th yr after grant	RM 1 400	RM 1 200
Annual fees for 18th yr after grant	RM 1 600	RM 1 300
Annual fees for 19th yr after grant	RM 1 800	RM 1 400
Annual fees for 20th yr after grant	RM 2 000	RM 1 500
Total costs	RM 16 700	RM 12 500

[12] See Section 19 as modified by the Second Schedule.

9.1.2 Examples of Granted Claims

As mentioned above, in the case of a utility innovation only one claim is allowed.[13]

Example 1:
'A composite pile (10) comprising a steel cylinder (12) plurality of steel bars (20) and a concrete portion (14) characterised in that the concrete portion (14) extends the entire length of the longitudinal steel cylinder and is attached to the inner surface of the said steel cylinder; and the plurality of steel bars being spaced apart within the concrete portion disposed at least at one terminal end of the pile or the entire length of the concrete portion, and wherein the unset concrete portion is spun after introduction of the unset concrete into the steel cylinder.'

Example 2:
'A heat reclaiming system for household refrigerator comprising a refrigerating circuit having a compressor (7), condenser coil (1), filter-drier (8), expansion device (9) and an evaporator, characterised by the fact that a liquid receiver (11) is connected between the outlet of said compressor coil and the inlet of said filter-drier; a heat exchanger (3) connected to receive refrigerant from the said compressor and to discharge refrigerant to the said compressor coil; the said heat exchanger is immersed in a reservoir of water contained in water tank (4), so as to obtain hot water by reclaiming the normally wasted heat rejected by the said household refrigerator and at the same time ensure efficient cooling of the said household refrigerator regardless of the changing water temperature in the vicinity of the said heat exchanger.'

Example 3:
'A camera (1) comprising:

- a camera body, said camera body being substantially rectangular;
- a lens (3), provided on said camera body for capturing an image on film;
- a viewfinder (2) for viewing the image to be captured through said lens (3) on film, said viewfinder (2) being located at a position near a bottom surface of the camera (1) and spaced from a top surface of the camera when the camera is oriented for horizontal format picture taking for view-

[13] This restriction is not specifically stated but is derived from Section 28(1)(d), as modified by the Second Schedule, of the Act.

ing through said viewfinder (2) without said camera body pressing against the photographer's nose (10).'

Example 4:
'An antimicrobial composition which comprises dried non-pathogenic micro-organisms in combination with an antimicrobial material derived from the plant family Allium, said antimicrobial material comprising alliin and being substantially free of allicin.'

Example 5:
'A process for producing solution of polyaluminium salt characterised by utilising the heat energy required for such reaction from the exothermic reaction of temperature range from 70 degrees C to 120 degrees C for a period of 1.5 to 3 hours, from its raw materials, sulfuric acid of 70 percent concentration, and bauxite, and reacting phospic acid of 85 percent concentration, to the resulting hot basic aluminium solution, adding soluble chloride, particularly calcium chloride or hydrochloric acid prior to or at the same time as calcium carbonate and basifying reaction until the evolution of carbon dioxide has ceased before cooling is to be introduced and to separate the sulfate ions as insoluble salts.'

Example 6:
'A chemical agent comprising phenoxyalkanoic acid, or a salt or an ester thereof for the control of unwanted plant species characterised in that the chemical agent consists of soluble amine salts of phenoxyalkanoic acid and is solid at ambient temperature and in a form that is introduceable into a spray tank to form a biologically active concentration of herbicide.'

Example 7:
'A shade cloth composed of warps (13) and wefts (11) having defined interstices (16) between the intersection of the warps and wefts characterised in that the warps and wefts are each formed from a multiplicity of discrete filaments (10) and wherein each weft (11) is composed of a continuous series of figure of eight loops (15) which interlace with the warps (13).'

9.1.3 Empirical and Policy Analysis of the Malaysian System

According to the Malaysian Intellectual Property Office, between 1986 and 2003, there were 1222 utility innovation applications. Table 9.2 represents the breakdown by year and field of technology.

The main users of the utility innovation system come from the region, with 47.3 percent of users coming from Taiwan, followed by 38.9 percent of the applications emanating from Malaysia. The third and fourth most popular users of

Table 9.2 Malaysian utility innovations based on field of technology

Section	88–89	'90	'91	'92	'93–95	'96	'97	'98	'99	'00	'01	'02	'03	Total	%
Human necessities	–	2	6	2	–	27	11	5	6	2	2	2	3	68	24.9
Performing operations; Transportation	–	4	5	1	–	30	8	3	3	1	2	1	3	61	22.3
Chemistry; Metallurgy	–	–	–	–	–	8	–	2	1	3	–	1	–	15	5.5
Textiles; Paper	–	–	–	–	–	3	2	–	–	–	–	–	–	5	1.8
Fixed constructions	–	3	5	–	–	10	4	1	2	2	4	–	–	31	11.4
Mechanical engineering; Lighting Heating Blasting Weapons	–	–	3	–	–	14	6	1	2	1	4	7	1	39	14.3
Physics	–	–	1	–	–	6	1	–	3	1	3	1	2	18	6.6
Electricity	–	–	2	–	–	12	5	–	6	2	7	2	–	36	13.2
Total	0	9	22	3	0	110	37	12	23	12	22	14	9	273	100.0

the system originate from the United States (4.3 percent) and Japan (1.3 percent). Turning to the proportion of utility innovation applications coming from companies and individuals and comparing the numbers with patents, one finds that from 1999 to 2003, 34.2 percent of the utility innovation applications came from companies and institutions, while 65.8 percent came from individuals. When one turns to the patent statistics during the same period, the percentages are very different: companies and institutions are responsible for 96.2 percent of applications with only 3.8 percent coming from individuals. It might appear from this that individual inventors find the utility innovation system to be more accessible than the patent system. Nonetheless, the total of utility innovation applications from individuals is substantially lower than the number of patent applications from individuals: 329 for utility innovations and 1102 for patents.

From this data, it can be concluded that the utility innovation system does have a place in the intellectual property system of Malaysia. The percentage of Malaysian innovators who make use of the system is high compared to the use of the patent system. The fact that the biggest users of the utility innovation system are from Taiwan could be explained by the fact that a lot of investors for the small and medium enterprises come from Taiwan.

Table 9.2 shows that the highest numbers of utility innovations encompass innovations relating to human necessities such as footwear, furniture, agriculture, jewellery and travelling articles. The second highest category is in relation to performing operations and transporting, and classifies innovations relating to mechanical operations involving physical or chemical processes, machines, apparatus and also transportation such as railways, aircraft and vehicles. These are areas in which individual innovators and SMEs could be involved in the creation of incremental improvements without the use of high technology. This may reflect the fact that the percentage of individual innovators compared to companies/institutions making use of the utility innovation system is very much higher as compared to the use of the patent system.

9.2 VIETNAM

French colonisation of Vietnam between 1867 and 1954 left an early patent system which was introduced in 1893. The patent system, in common with other colonised countries, was basically a re-registration system with patents being issued in France and being made applicable in Vietnam. After 1954, South Vietnam retained a patent regime, whereas North Vietnam did not have an intellectual property system. The Socialist Republic of Vietnam began to protect industrial property in 1981, nearly six years after unification when it promulgated the Regulations on Innovations and Inventions by Decree No. 31-CP. The law conferred inventors' certificates and patents. Subsequently, the Regulations on Trademarks dealing with the registration and protection of trademarks were issued on 21 December 1982. UM protection was first introduced by the Regulations on Utility Solutions (Decree No. 200/HDBT of 28 December 1988). All these regulations have been replaced by the Civil Code. Thus, the current law on intellectual property is the Decree on Industrial Property No. 63-CP effective 24 October 1996, pursuant to Civil Code dated 28 October 1995, Part 6.

9.2.1 Substantive Law

According to the Civil Code, protection is accorded to 'inventions, utility solutions, industrial designs, trade marks ...'. A 'utility solution' is defined as a technical solution which is new on a worldwide basis and applicable in socio-economic areas. A patent for an invention, on the other hand, is defined as a technical solution, which is new on a worldwide basis, involves an inventive step and is applicable in socio-economic areas.[14] The following are statutorily excluded from protection: (i) signs, diagrams and symbols (ii) computer programs and layout designs of integrated circuits (iii) plant or animal varieties (iv) processes of prevention, diagnosis of diseases and treatment of human beings, animals or plants.

As can be seen, novelty is universal for both major and minor innovations; the latter, however, do not require an inventive step. The criterion of social and economic application is related to industrial applicability.[15] The criterion of novelty for both invention and utility solution is identical. A utility solution is considered new if:

i. prior to the filing date or, where appropriate, the priority date of the patent application, the invention or utility solution was not described in the earlier domestic patent applications for invention and/or the earlier domestic patent applications for utility solution
ii. the invention or utility solution has not been disclosed in Vietnam and/or abroad in such a way as to permit the realisation of that invention or utility solution.

The disclosure of an invention or a utility solution such that it becomes part of the state of the art may take place in three ways, namely (i) publications in tangible form (ii) oral disclosure and (iii) disclosure by use.

---

[14] Sections 782 and 783, Vietnamese Civil Code.

[15] Heath, C. (2003), 'Vietnam', in *Intellectual Property Law in Asia* (ed. Christopher Heath), Kluwer Law International/Max Planck Series on Asian Intellectual Property Law, Volume 5, p. 274.

To meet the formal requirements, an application must be filed on an official form. The examination as to formal conditions must be completed within three months from the filing date. All national patent and utility solution applications are published for purposes of opposition 18 months after the priority date unless a request for an earlier publication is filed or a request for examination as to substance is filed before 18 months from the priority date. The default system is that the Vietnamese patent law operates a deferred examination system for patents and UMs. Rather than *ex officio*, substantive examination will be granted upon an applicant's request. In the case of a patent, the request must be made not later than 42 months from the priority date, and in the case of the UM, the period is 36 months. The current law provides that any time during the substantial examination, any third party may request rejection of a patent or utility solution application by submitting arguments together with relevant documents and information with respect to the patent application.

The law offers the same rights to both types of patent holders. Any unauthorised person who:

- makes, uses, sells, offers for sale, stocks for sale, imports or advertises the patented product or
- uses a patented process or makes, uses, sells, offers for sale, stocks for sale, imports or advertises the product directly obtained through the patented process

will be deemed to have violated the exclusive rights of the owner of the patent.

9.2.2 Empirical and Policy Implications

The trend under the Vietnamese system shows a steady rate of applications on the whole but reveals a drop of applications from local Vietnamese, from 52 in 1991 to an average of 30 from 1999 to 2001. The rate of applications by foreigners has, however, increased. Applications have risen from an average of one to a total of 58 in 2000. The National Office's filing statistics show that the top countries from which foreign applications for *both* patents and utility solutions originate from are (the figures in parentheses are 2000 and 2001 application figures):

1. United States (437 349)
2. Japan (218 217)
3. Germany (90 129)
4. China (2986)
5. France (8869)

Table 9.3 Utility solution applications from 1989 to 2001

	Applications by Vietnamese	Applications by foreigners	Total applications
1989	25	0	25
1990	39	25	64
1991	52	1	53
1992	32	1	33
1993	38	20	58
1994	34	24	58
1995	26	39	65
1996	41	38	79
1997	24	42	66
1998	15	13	28
1999	28	14	42
2000	35	58	93
2001	35	47	82

Source: WIPO and National Office of Industrial Property of Vietnam.

Table 9.4 Utility solution grants

	Grants to Vietnamese	Grants to foreigners	Total grants
1990	23	0	23
1991	44	1	45
1992	23	1	24
1993	9	1	10
1994	8	9	17
1995	18	16	34
1996	5	6	11
1997	8	12	20
1998	3	14	17
1999	6	12	18
2000	10	13	23
2001	17	9	26

Source: National Office of Industrial Property of Vietnam.

6. United Kingdom (5962)
7. Switzerland (5162)
8. Netherlands (6552)

9.3 THAILAND

The first law to protect intellectual property rights in Thailand was the Copyright Act of 1892. Despite this early start in respect of literary and artistic works, no law to protect inventions and innovations emerged until 1979. Indeed, the Thai Supreme Court had held in 1964 that rights over inventions or patents were not proper legal rights. Unlike other ASEAN counterparts who had this regime imposed by colonial rule, the Thai patent law emerged from economic policy and was an attempt to accelerate the R&D capacity of the country. The Patent Act B.E. 2522 (1979), was subsequently amended by the Patent Act B.E. 2535 (1992) – this act was amended partly from pressure by the United States government and in an attempt to avoid trade sanctions. Another revision took place, this time to comply with the TRIPS Agreement. This latest Act entered into force on 27 September 1999. One of the most significant changes is the introduction of 'petty patents'.

9.3.1 Substantive Law

Thai patent law covers three kinds of patents: patents for inventions, design patents and petty patents. A petty patent is a document which is granted to an invention or a design.[16] In turn, an invention is defined as any innovation or invention which creates a new product or process, or any improvement of a known product or process. The definition also includes any improvement of a known product or process. Natural occurring microorganisms and their components; animals, plants or extracts for animals or plants; scientific and mathematical rules or theories; computer programs; methods of diagnosis, treatment, cures of human diseases; and inventions contrary to public order, morality, health or welfare are statutorily excluded from petty patent protection.

To be eligible for petty patent protection, an invention must be new and industrially applicable, though it will not need to possess an inventive step. An invention is new if it does not form part of the state of the art. The state of the art includes any invention which is widely known, used, disclosed or applied for anywhere in the world before the application date. It is possible to conduct a pre-filing novelty search to ensure that the petty patent is new and that a post-

---

16 S. 65(a), Thai Patent Act.

grant challenge to the validity of the patent in respect of novelty will fail. Under the Patent Act an invention is deemed to be industrially applicable if it can be made or used in any kind of industry including handicrafts, agriculture and commerce. Under the Thai patent law the same invention cannot be the subject of both a petty patent and a patent. The applicant for a petty patent or an invention patent may request to convert his application for a petty patent to an application for a patent or vice versa before the publication of the application.

In order to get petty patent protection, an application form containing the title, features and purpose and use of the creation must be filed with the Department of Intellectual Property. Petty patent applications are subject to a preliminary examination of formalities. This examination also checks whether the technical creation belongs to one of the statutorily excluded categories for which protection cannot be granted.

Petty patent applications are not examined automatically, and a request can be made for a substantive examination within five years from the date of publication. The risks of this lack of substantive examination are, first of all, that it is possible that several UM applications are granted for identical inventions and also that the petty patent granted conflicts with earlier patented rights. However, the patentability of a petty patent registration can be challenged. Any interested person can file a request with the Department of Industrial Property (DIP) within one year of the publication date. A competent official will conduct the examination of the patentability of the petty patent and will produce a report which will be submitted to the Director General of the DIP. The Director General's decision concerning the validity of a UM can be challenged in court. The owner of a petty patent has the right to use the words 'Thai petty patent', its abbreviation or any foreign word of the same meaning on the product, the packaging or the advertisement of the product.

9.3.2 Empirical and Policy Implications

Table 9.5 Utility model applications from 1999 to 2004

	Applications by Thai	Applications by foreigners
1999	185	17
2000	555	67
2001	745	66
2002	1148	74
2003	1290	54
2004	1390	64

Source: Thai Department of Industrial Property.

Table 9.6 Utility model applications from 1999 to 2004

	Grants to Thai	Grants to foreigners
1999	7	0
2000	108	17
2001	341	51
2002	376	13
2003	476	11
2004	364	28

Source: Thai Department of Industrial Property.

9.4 INDONESIA

The first intellectual property laws in Indonesia were introduced in 1844 by the Dutch East Indies in respect of inventions. This law granted royal privileges but was abolished by 1870. No further laws on patents were introduced until the Dutch legislation of 1910. In common with other colonial countries, it was merely a re-registration system. After independence, the first modern Indonesian patent law was introduced in 1989, to be replaced by the current 2001 Patent Act. The UM system in Indonesia, called the petty patent system, is governed by the Patents Act No. 14 (2001).

A petty patent is an invention which is defined as 'the idea of an inventor moulded into an activity to solve a specific problem in the field of technology which may take the form of a product or a process, or of the perfection and development of a product or a process'.[17] Excluded from patent protection are the following:

- inventions concerning processes or products whose publication and use or implementation is in conflict with effective legal provisions, religious morals, public security or ethics;
- inventions that involve examination, treatment, therapy and/or surgical operations for people and/or animals;
- inventions on theories and methods in the field of science and mathematics;
- inventions concerning living creatures of any kind with the exception of microorganisms;

[17] Section 1, No. 2 Patents Act 2001.

- inventions concerning biological process essential for the production of plants or animals with the exception of non-biological processes and microbiological processes.

The invention must be new and industrially applicable but does not need to show inventiveness. The test for novelty is identical for both standard and petty patents i.e. the invention is deemed novel if it has not been published inside or outside Indonesia prior to the application date or the priority date, in writing, orally or by way of demonstration or in any other manner that would enable an expert to implement the invention.[18] An invention is industrially applicable if it can be implemented in the industry as described in the application. An examination of formalities of the application is conducted by the Directorate General of Intellectual Property Rights. A request for a substantive examination on a petty patent may be filed at the same time as the filing of the application or at the latest, six months from the filing date prior to payment of a fee. In conducting the substantive examination, the Directorate General will only examine the novelty and the industrial applicability of the invention. The patentee has the exclusive right to produce, use, sell, import, lease or transfer the protected product or make it available for sale, lease or transfer. In relation to a process patent, the patentee has the exclusive right to use the patented production process for the activities stated. There is a limitation to the rights which states that patentee's rights do not extend to any act made in the interest of education, research and long term analysis, if such use does not harm the legitimate interests of the patent holder.

9.5 PHILIPPINES

A protection regime for intellectual property has been in existence in the Philippines since the Spanish occupation prior to 1862. Patent protection was available from the mid nineteenth century with patents being granted to inventors for a term of 20 years. After the cession of the Philippines to the United States, a 1913 Act provided for the applicability of United States patent laws in the Philippines; moreover, a re-registration system commenced, with patent applications from the Philippines being filed with the United States Patent Office. United States laws were repealed after the Second World War, and the governing law today is the Intellectual Property Code (Republic Act No. 8293, 1 January 1998).

---

[18] Section 3(2), Patents Act 2001.

9.5.1 Substantive Law

Any technical solution to a problem in any field of human activity which is new and industrially applicable will qualify for registration as a UM. In particular, a UM may be, or may relate to, a useful machine, an implement or tool, a product or composition, or an improvement of any of the foregoing. An invention qualifies for registration as a UM if it is new and industrially applicable. Mere local novelty is sufficient.

The UM application is examined only as to formalities. Nevertheless, to assist the applicant in considering his options under the IP Code (for example, patent protection) the application will be classified and a search conducted to determine prior art. The registrability report will contain citations of relevant prior art documents with appropriate indications as to their degree of relevance which will serve as an aid to the applicant in the determination of the validity of the UM claim(s) in respect to newness.

Table 9.7 Utility model applications from 1983 to 2001

	Applications by Filipinos	Applications by foreigners	Total applications
1983	409	27	436
1984	474	2	476
1985	449	13	462
1986	346	6	352
1987	354	8	362
1988	330	7	337
1989	367	32	399
1990	321	16	337
1991	456	18	474
1992	504	24	528
1993	499	44	543
1994	605	34	639
1995	642	40	682
1996	555	47	602
1997	517	66	583
1998	602	31	633
1999	606	41	647
2000	536	21	557
2001	429	21	450

Source: Intellectual Property Office of Philippines.

9.5.2 Empirical and Policy Implications

Table 9.8 Utility model grants from 1983 to 2001

	Grants to Filipinos	Grants to foreigners	Total grants
1983	288	8	296
1984	235	2	237
1985	294	8	302
1986	278	8	286
1987	275	7	282
1988	192	5	197
1989	192	4	196
1990	198	3	201
1991	170	7	177
1992	145	8	153
1993	260	3	263
1994	127	8	135
1995	221	3	224
1996	177	10	187
1997	263	8	271
1998	3	0	3
1999	314	3	317
2000	287	1	288
2001	189	5	194

Source: Intellectual Property Office of Philippines.

The registration rates in the Philippines, in contrast to Vietnam, show that the system is robustly utilised by the local innovators, as opposed to non-residents. Residents make an average of 500–600 applications per annum as opposed to non-residents who make an average of 30 applications per annum. Contrast this to the Vietnamese situation where applications by residents and non-residents average to approximately similar figures.

9.6 CAMBODIA

Cambodia introduced a UM law under its Law on the Protection of Patents, Utility Model Certificates and Industrial Designs, which was adopted by the Council of Ministers on 12 October 2001. With the exclusion of a few provisions, the patent provisions apply *mutatis mutandis* to UM certificates. A UM certificate is defined as 'any invention which is new and industrially applicable and may be or may relate to a product or process'. The term 'invention' is also statutorily defined as 'an idea of an inventor which permits in practice the solution of a specific problem in the field of technology'. The law states that where the commercial exploitation of inventions is contrary to public order or morality, such inventions will be excluded from patent and UM protection.

As just stated, the criteria of protection are novelty and industrial applicability. An invention is deemed to be new if it is not anticipated by any prior art. Prior art consists of everything disclosed to the public anywhere in the world by any tangible or oral disclosure or use prior to the filing date. An invention shall be considered industrially applicable if it can be made or used in any kind of industry, with 'industry' being defined in its broadest sense.

The UM certificate expires, without any possibility of renewal, at the end of the seventh year after the filing date. At any time before the grant or refusal of a patent, an applicant for a patent may, upon payment of the prescribed fee, convert his application into an application for a UM certificate, which shall be accorded the filing date of the initial application. The same applies vice versa.

9.7 CONCLUSION

The evidence shows that the usage of UMs is not high in many ASEAN countries. Vietnam, for instance, shows an annual average application rate of less than 100 applications whereas the Philippines has an average application rate of 400–500. What are we to make of these statistics? Evidence clearly does show that UMs are of more benefit to local innovators than international firms. The Taiwanese regime indicates that UMs are incredibly popular with the local innovators, with a low degree of invalidity. Nevertheless, these figures do not answer the question as to whether the Philippines has a greater innovative climate than Vietnam or Singapore (without a UM system) merely because of high application statistics.

A recent newspaper article (the *Singapore Straits Times*) pointed out that productivity gains from new knowledge and technological progress contributed just a third or less to the increase in income within the ASEAN region. In contrast, productivity gains were the source of four-fifths of income growth in the United States, and two-thirds in France, Germany and the United Kingdom during the same period. The report stated that ASEAN as a whole needed to move up the innovation ladder. Apparently, fewer than 1,600 invention patents had been granted to ASEAN residents at national intellectual property offices from 1993 to 2002. This number represented less than one percent of all the ASEAN patents granted by these offices to residents and foreign inventors. Intriguingly, the one exception within ASEAN is probably Singapore which accounts for nearly 60 percent of patenting activity within the ASEAN region.[19] Indeed, Singapore is atypical also in the sense that as much as 72 percent of United States patents issued to ASEAN residents belonged to Singaporeans.[20]

[19] Wattanapruttipaisan, T. (2004), 'Innovation is ASEAN's Achilles Heel', *The Straits Times*, 30 November.

[20] http://www.aseansec.org/17304.htm.

10. Utility models in Latin America

Manuel Marquez

10.1 UTILITY MODELS IN LATIN AMERICA

In Latin America, examples exist of regional utility model regimes that hold each country to harmonised standards (the Andean Community) and of countries designing their own national system (that is Mexico, Argentina and Brazil). This chapter covers the Andean Community Common Regime on Industrial Property and the Mexican UM regime.

The Andean Community

Andean Community Decision 486 of 2000 established a Common Regime on Industrial Property for the five member countries (Bolivia, Colombia, Ecuador, Peru and Venezuela). The decision does not create unitary IP regimes but defines standards that must be adopted by each member.

Title III of the decision deals with UMs. A UM is considered to be a 'new form, configuration, disposition of elements, of any artefact, tool, instrument, mechanism or other object or any part of the same, that permits a better or different functioning, use or manufacture of the object which incorporates or which offers any use, advantage or technical effect that it did not have previously'. The term of protection is ten years from the filing date. UM applications can be converted upon request to patent applications as well as vice versa.

10.2 A CASE STUDY ON MEXICO

Mexico, on the other hand, opted to design its own system. Efforts to legally protect technological innovation have a long history in that country. Indeed, the first regulation for the protection of inventions was enacted in 1820, when Mexico was the Spanish colony of New Spain. Since those days, legal administration in Mexico has been very active in this respect. Mexico has passed industrial property laws or revised them 19 times from that year to 2005, changing terms of protection, patentability requirements, office procedures, fees, and penalties for rights infringements.

Two very important changes took place in 1976 with the establishment of an inventor certificate (IC) regime, and in 1991 with the provision of UMs.[1] For the patent and IC regimes, the International Patent Classification (IPC) was adopted from 1979 for their classification and publication. ICs remained available until 1991 when they were abolished.

In the mid 1990s, two other important developments took place. In 1994, the Mexican Patent Office was renamed the Instituto Nacional de Propiedad Intelectual (IMPI) and turned into an independent organisation with administrative control over its own resources. In 1995, Mexico signed the Patent Cooperation Treaty (PCT). This has led to a rise in the number of applications received from nationals of state party countries.

10.2.1 Trends in Patenting

From 1890 to 2005, a total of 228,815 patents have been granted. Since 1995, the number has risen markedly. There are a number of possible explanations for the increase. Since 1985, Mexico has signed several free trade agreements including with 15 countries and the EU. Mexico also joined the World Trade Organization and was thus obliged to implement the TRIPS Agreement. This period is recognised as 'the commercial opening of Mexico', which brought as a consequence the need to modify the judicial framework for the registration and protection of technology. Also, in 1986, the period of protection was extended from 10 to 14 years from grant, and in 1991 to 20 years from the filing date. In the same year, the IC system was abolished. During that period, also, the patent system became available in more technological fields. But the most significant influence was Mexico's ratification of the PCT.

However, while the number granted to Mexican nationals has increased slightly, as a proportion of the total, it has declined, falling to only about 2 percent (see Table 10.1).

10.2.2 Promoting Innovation Without Patents

Inventor certificates

When ICs were created as an alternative to patents for technology protection, some areas of technology were deliberately excluded from patents and placed within the IC regime. These included 'procedures to obtain chemical products,

[1] Industrial Property Law (of 25 June 1991, as last amended by the Decree of 17 May 1999).

Table 10.1 New patent grants awarded to Mexican nationals, 1995–2004

Year	New grants	Grants to nationals
1995	3 537	148
1996	3 186	116
1997	3 946	112
1998	3 219	141
1999	3 899	120
2000	5 519	118
2001	2 479	118
2002	6 611	139
2003	6 008	121
2004	6 835	162

alloys and pharmaceutical chemical processes; inventions related to nuclear energy; apparatus and equipment for pollution and its manufacturing process, modifications or applications'.

As with patents, applications were made to the patent office. The only difference is that ICs were cheaper to obtain. The protection term was identical: ten years from the date of grant. The coexistence of the patent and IC regimes was intended to enhance local innovation and acquisition of technology. It was thought that the IC system would be used mainly by local people. For one thing, the extent of exclusivity for right holders was less than with patents. And for another, the Transfer of Technology Office (Ministry of Commerce) had the power, albeit never exercised, to intercede in the case of refusal to license and establish a fixed royalty rate. Shifting technological fields from the patent system to the IC regime, it was expected, would result in fewer patent grants to foreign companies. However, from 1976 to 1991, out of 7734 ICs registered, just under 3 percent were from Mexican nationals, so the IC regime failed in its basic aims.

Utility models

The Mexican Industrial Property Law defines a UM as follows:

> Art. 28. – …the objects, devices, apparatus or tools which, as a result of a modification in its arrangement, configuration, structure or shape, present a different function with respect to the parts which complete it or advantages in its utility.

As can be seen, the UM is not for the invention of a thing, but for making existing things work better. The period of protection is ten years from the filing date.

The application procedure is similar to a patent. The applicant has to prepare a description, mentioning the technical background, the objective of the invention and describing the invention. The explanation can be supported by drawings. Finally claims have to be mentioned and a summary of the invention. Also, the UM must be for a single invention.

In Mexico, neither the law nor the courts contemplate that UMs should be treated any differently from patents. However, UM applications are not published, and the public is not notified in any way about their existence until after they are granted. Patents, on the other hand, are published 18 months after filing. It is important to mention that the filing of UM applications costs about half that of patents. Concerning the possibility of converting a UM to a patent, the rule is that if the technical subject matter justifies it, an examiner can suggest that a UM application be changed to a patent application. However, the fees payable would be that for a patent, not a UM. Conversely, a patent application deemed not to qualify may be changed into a UM application.

Applications for UM protection were accepted from 1991, and by 30 June 2005, 1341 UMs had been granted. Of these, 1041 (78 percent) were granted to Mexican nationals (see Table 10.2). Although much fewer UMs have been granted than ICs during a similar period of time, for policymakers seeking to encourage domestic usage of intellectual property law, the UM system appears to function more effectively. Nonetheless, the amount of UMs granted remains modest, and with such low granting rates compared to patents (and also with ICs), the regime can be considered so far to be no more than a qualified success if a success at all.

Table 10.2 New utility model grants awarded to Mexican nationals, 1992–2000

Year	New grants	Grants to nationals
1992	39	30
1993	97	78
1994	143	94
1995	221	159
1996	20	16
1997	64	54
1998	83	68
1999	90	61
2000	104	69

10.3 CONCLUSIONS

Twenty years ago the commercial opening of Mexico began as the country sought to improve its economic conditions and international competitiveness. Consequently, commercial treaties were signed with several countries, the EU, Japan and other various international organisations. Another response by Mexico was to modify its Industrial Property Law to enhance incentives for domestic intellectual property ownership over new technologies. However, foreign businesses were often better placed to take advantage of the reforms. Thus, while the number of patent grants increased 100 percent, from 3500 in 1995 to almost 7000 in 2004, the number of national grants held steady, around 100–150 per year.

In contrast, while UM grants have not yet experienced such overall increases, national grants are proportionately very high. The question for policymakers to ask themselves is whether it is more important for Mexicans to engage significantly in an IP regime that has modest overall participation levels than for them to be far less involved in an albeit more popular system.

PART THREE

Diverse paths to an innovative future

11. Conclusions and recommendations

Uma Suthersanen and Graham Dutfield

11.1 JUSTIFYING IPRs

Patents are tools for economic advancement that should contribute to the enrichment of society through (i) the widest possible availability of new and useful goods, services and technical information that derive from inventive activity, and (ii) the highest possible level of economic activity based on the production, circulation and further development of such goods, services and information. In pursuit of these aims, inventors are able to protect their inventions through a system of property rights – the patent system. One of the reasons that patents are so controversial is that the IP incentive, as far as it actually works, functions by restricting use by others of the protected invention for a certain period. Yet follow-on innovation by others is more likely to happen if use is not restricted. Thus a balance between private control over the use of technical information and its diffusion needs to be struck. Where the line should be drawn is very difficult to determine but its ideal location will vary widely from one country to another, and, one may argue from one business sector to another. In countries where little inventive activity takes place, free access to technical information may well do more to foster technological capacity building than providing strong private rights over such information.

We face two conflicting schools of thought. An economic classicist would argue that the main role of patent law is to prevent rewards from being dissipated by competing imitations and to preserve the incentive for continual innovation leading to economic growth. If we unconditionally accept this perspective, the rationale for increasing the scope of patent protection to encompass minor or incremental inventions is clear.

Nevertheless, we are also cautioned against broadening protection to such an extent that the excessive allocation of rights to an innovator will lead to a severe drop in the number of competitive substitutes or may completely prevent any other competitive substitute from appearing in that market.[1] From this perspec-

[1] Scherer, F.M. and D. Ross, *op. cit.*, pp. 577–8.

tive, we are reminded that, considering the heavy price that the society will have to pay in stifling competition in respect of innovations, patent law has been designed to only confer exclusive proprietary rights in certain types of inventions *viz.* those inventions which exhibit some technical character, and which manifest the necessary level of inventiveness and novelty. A more forceful objection to minor inventions is made by Kitch who states that 'low cost inventions' should not be patentable since the inventor is always rewarded by a head start, and these types of inventions would exist anyway in the absence of a patent system.[2]

How then do we justify lowering the thresholds of protection in order to extend protection to minor or incremental innovations? The argument that market failure will result if excessive protection is granted to inventions cannot be easily deflected by reasons of administrative or cost problems, or by the fact that some inventions are susceptible to unfair copying. Nonetheless, introducing a utility model system could potentially stimulate further innovation. If so, it would do so on the following bases:

i. Less knowledge-based[3] industries would be able to seek protection for innovations which do not meet the necessary requirements or need the level of protection required for a standard patent.
ii. High knowledge industries like semiconductors, ICT-related products and computer peripherals would be able to protect lower-end innovations with a more economical alternative to patents.

The following questions then arise for policymakers of developing countries without UM regimes:

1. Does the country in question require rapid, cheap protection for local industrial growth and promotion?
2. Is there an economic argument as to why such inventions should be protected?
3. Are the current patent and design regimes suited to industry's needs in terms of criteria of protection, costs and ease of use?
4. Is there a need to reconsider the economic and legal policy of granting patent protection to innovative activities; and if there is, whether these new policies are best served within a one-tier or two-tier patent regime in the developing country (or even a three-tier system if one includes the design right regime)?

---

2 Kitch, E. (1977), 'The nature and function of the patent system', *Journal of Law and Economics*, **20** (265), p. 281.

3 Apparently, the major users of the Australian innovation patent system are from the less knowledge-based industries as opposed to the knowledge-intensive users.

5. Is there massive copying of minor innovations?
6. Is there a need to maintain a large public domain to assist follow-on innovation?
7. Would the country be placed at a disadvantage if other countries use a utility model system in a major way?

11.2 POLICY OPTIONS FOR DEVELOPING COUNTRIES

European industries have complained that they are now vulnerable to unfair copying by competitors and that the unavailability of patent protection robs them of the vital lead-time required to recoup research and developmental costs. Nevertheless, industry has also opposed a European Community-wide UM regime out of fear of restricting competition. If we look at the patterns of inventive activity in Europe, it is not at all clear that UMs are vital for innovative growth – witness Sweden's top position in the European innovation scoreboard and its lack of a second tier patent system. Indeed, a more interesting development is the European attempt to broaden the scope of patent law to encompass more subject matter and offer wider scopes of protection – this may, following the United States, be the approach to adopt.

It might be that rather than follow the experiences of countries already having UM systems, developing countries without UMs would be better off with a policy of 'intellectual property leapfrogging'. By this, we mean that they should consider eschewing the evolutionary approach adopted by these countries (and many of today's rich countries in the past) of opting for UMs as a means to accelerate their advance to developed country status, and instead learn from and emulate present-day Japanese and United States innovation promotion regulation where the role of UMs is either diminishing or was never existent.

It is advisable to determine the long term sustainability of the UM system with a cost-benefit analysis, weighing carefully the pros and cons and their possible economic impact to which this study acts as a precursor. In implementing a UM system, careful planning and effective dissemination of information, as well as developing the awareness of the target market would be of critical importance to the success of the new system.

There are three options which a developing country can consider.

1. Status Quo Approach

A developing country can accept the existing intellectual property regime, without introducing any new right.

This may be due to the fact that an analysis of the country's economic and industrial environment fails to demonstrate that the country stands significantly to benefit in terms of innovation or economically from introducing a UM system. Note that there is always some unhappiness from the local stakeholders' perspectives as to the cost, length of time before rights are granted, and about the complexity of the current intellectual property system. If incremental inventions are to be protected, there must be a valid reason as to why a UM system is to be introduced, and why this proposed law will institute a more favourable regime as compared with the current patent or design laws.

2. Accretion Approach

A developing country can adjust the existing intellectual property regime without introducing a UM right. This can be done by extending existing intellectual property rights to new subject matter (such as sub-patentable or functional innovation) by re-defining an existing right to encompass the new material.

The accretion method involves a minor tinkering of the country's patent or design or unfair competition system.[4] In relation to the patent system, it is worthwhile to note the reasons suggested for the high patenting figures in the United States including a decreased invalidation rate and the commensurate pressure on the United States Patent and Trademark Office to grant patents.

Another point worth considering is the ability of design law to encompass very wide subject matter – indeed, the TRIPS Agreement is incredibly flexible in relation to the criteria and scope of protection in relation to designs. Moreover, design law has been the historical basis of protection of minor innovation in both the United Kingdom and Germany. Nevertheless, it should be recalled that the design laws in Australia and Germany were considered too narrow to protect functional innovations and hence UM laws were introduced. In the United Kingdom, an alternative solution was found to protect minor functional innovations – an unregistered design law was introduced. Finally, is there room for a misappropriation tort or a more extensive passing off or unfair competition law? This alternative has been adopted by Japan which has special 3–5 year unfair competition laws to protect minor innovation.

3. The Emulation Approach

Emulation involves creating new hybrid rights. If a developing country does not have such a right, it would be the most expensive option in the short run. However, this expense is an immediate real cost which may be offset by long term

---

4 Cornish, W.R. (1993) 'The international relations of intellectual property', **52** (1) *Cambridge Law Journal* 46, pp. 54–5.

benefits to the industrial environment such as increased international licensing opportunities.

11.3 THE IDEAL LAW FOR MINOR INNOVATIONS?

Should it still be felt that policy considerations necessitate the introduction of a system in a particular country for minor innovations in the form of a property right, it is recommended that the following essential features be considered:

- **Subject matter of protection**: the law should comprise a detailed list of excluded subject matter which must mirror the exclusions under the patent law. Moreover, it is worth considering excluding some types of invention as dictated by public policy such as chemicals or pharmaceuticals or biological material or substances or processes.
- **A renewal-based term of protection**: the term should be a minimum term of three years (as the absolute minimum) but not exceeding the patent term of 20 years. However, a more appropriate maximum term based on existing UM and design laws in several jurisdictions is ten years.
- **A non-examination system**: at least for the first period of registration. This is one of the key ingredients for the encouragement of innovation.
- **A compulsory examination/report for second stage**: this procedure could be introduced if the term of protection is linked to a renewal procedure.
- **A compulsory examination/report for invalidation/infringement proceedings**: this burden should be placed on the property rights holder to produce (and pay for) a detailed examination report as to the novelty and inventiveness of his innovation prior to litigation.
- **Renewal fees**: this is one suggested means to weed out the underutilised property right.
- **Tiered fees**: this is to enable the system to be friendly to small and medium sized enterprises or the individual inventor.
- **Novelty**: most developed countries have introduced universal novelty within the utility model system; whereas some countries which have protected technology or functional designs under either design or *sui generis* rights have introduced the 'commonplace in the relevant field' criterion (the United States Semiconductor Chip Protection Act 1984 and the United Kingdom unregistered design right). Moreover, enforcement problems develop if a lower level of novelty (such as domestic or regional novelty) is adopted.
- **Government action**: introducing a completely new intellectual property system requires some effort on the part of all the relevant stakeholders

(such as the relevant government agencies and patent attorneys/agents) to engender 'an IPR culture' such as an awareness education training programme, helpdesk services, web and print literature, innovation fairs etc. to inform potential users of how the new system works, and how it differs from existing standard patent and design systems.

- **Cross-licensing/compulsory licensing**: a final issue is whether compulsory licensing provisions should be included on the same basis as patent law. This is especially important in the case of a new UM law where two rights conflict and a subsequent inventor cannot obtain or exploit a patent or UM without infringing a prior patent or UM. This is a further consideration if a no-examination system is adopted.

Index

African Intellectual Property Organization (OAPI) 31
African Regional Industrial Property Organization (ARIPO) 35
Agenda 21 5
Andean Community 188
ASEAN 169, 186–7
  see also individual countries
Australia 119, 139–41
  design law 43, 198
  empirical analysis of second tier patent system 126–38
    in Australian patent context 127–8
    Australian *versus* foreign applicants 130–32
    country of origin of foreign applicants 132–6
    individual *versus* company applicants 128–30
    technology groups of applications 137–8
  innovation patents 18, 32, 139, 141
    empirical analysis 127–8, 131, 134–6, 138
    features 125–6
    objectives 125
  patents 119–20
  petty patents 32, 119, 126, 139–40
    empirical analysis 127–8, 131, 133–4, 137–8
    features 122–3
    history 120–21
    objectives 121–2
    review 123–5

bargaining tools 42
Belgium
  patents 25
  short term patent 18, 24
Bentham, Jeremy 13
biotechnology patents 93–6
Brandt Commission 3
Brundtland Commission 4–5

Cambodia
  utility models 34, 185–6
case studies
  Singapore 75, 110–17
    electronic systems company 113–15
    healthy lifestyle products company 110–13
    water treatment company 115–17
China 152
  design law 154
  economic and innovation climate 152–3
  patents 153–6
  utility models 32, 153, 154–6
    empirical analysis 156–62
    policy implications 162–3
commercialization 36–7
Community Design Right (CDR) 46–7
competition, unfair 43, 49–55
  interactions with other laws 56–7
copying
  benefits of 41
  as creative imitation or misappropriation 66–8
  innovation and 10–11
  Singapore 104
  unfair 35–6
  unfair competition 43, 49–55
    interactions with other laws 56–7
copyright
  Singapore 102, 103
Cornish, William 43
cumulative innovation 7

Denmark
  economic and innovation climate 22, 23
  utility models 26, 27

design law 24, 43, 44–8, 198
  Australia 43, 198
  China 154
  European Union (EU) 45–8, 55
  Germany 43, 198
  interactions with other laws 56–7
  Thailand 180
  United Kingdom 24, 43, 47, 48, 55, 198
  United States of America 44–5
developing countries 3–4, 5–6
  policy considerations 197–9
    accretion approach 198
    emulation approach 198–9
    status quo approach 197–8
discrete innovation 7

economic climate
  China 152–3
  European Union (EU) 22–3
  Germany 22, 23, 27–8
  Japan 22, 23, 142
  United States of America 22, 23
economic development 3–5
  innovation and 5–10
Ericsson 10
Ernst, D. 9
European Innovation Scoreboard 22
European Patent Convention 23
European Patent Office 24
European Union (EU)
  design law 45–8, 55
  economic and innovation climate 22–3
  patent classification 25–6
  three-dimensional classification 26–7
  unfair competition 52
  utility models 24–5, 40
  see also individual countries
exporting IP products 65–6

Finland
  economic and innovation climate 22, 23
  utility models 26, 27
France
  economic and innovation climate 23
  patents 25
  unfair competition 54–5
  utility certificate 18, 24, 25

Germany
  design law 43, 198
  economic and innovation climate 22, 23, 27–8
  intellectual property institutional order 65
  patents 27–8
  unfair competition 52–4
  utility models 28–31, 56, 65
    current law 28–30
    empirical evidence 30–31
Geroski, P. 14
Greece
  utility models 27
Guile, Bruce R. 36

ul Haq, Mahbub 4
Harare Protocol 35
Hong Kong
  utility models 32
Howard, W. G. 36
human development 4

ideal law for minor innovations 199–200
imitation *see* copying
importing IP products 65–6
Indonesia
  patents 182
  utility models 34, 182–3
information and communications technology patents 93–6
innovation
  copying and 10–11
  economic development and 5–10
  innovation climate
    China 152–3
    European Union (EU) 22–3
    Germany 22, 23, 27–8
    Japan 22, 23, 142
    United States of America 22, 23
  learning how to innovate 10–11
  minor innovations 17
    ideal law for minor innovations 199–200
  nature of innovator and innovation 68–9
  Singapore 83–8
    benchmarking 90–97
    international comparisons 88–90
    level of innovation 99

utility models and innovation culture 150–51
innovation patents *see* utility models
intellectual property rights
import/export of IP products 65–6
justification 195–7
right type of intellectual property institutional order 64–5
see also patents
International Patent Classification (IPC) 19, 21, 189
inventor certificates
Mexico 189–90
Ireland
economic and innovation climate 23
patents 25
Italy
economic and innovation climate 22
utility models 25, 27

Jaffe, Adam B. 59
Japan
economic and innovation climate 22, 23, 142
patents 42
historical background 142
Japanese and foreign applicants 147
numbers 146
utility models 32
comparison with Korea 148–50
correlation between UM and innovation culture 150–51
decrease of UM registrations 145–8
historical background 142–3
individual *versus* company applicants 147
Japanese and foreign applicants 148
procedure and examination 143–5
scope of protection 145
subject matter and criteria of protection 143
Jefferson, Thomas 14–15
Juma, C. 37

Kazakhstan
utility models 32
Kim, L. 11
Kitch, E. 196
Korea (South)
utility models 32, 148–50
correlation between UM and innovation culture 150–51
Kortum, Samuel 58–9
Kumar, N. 37
Kyrgyz Republic
utility models 32–3

learning how to innovate 10–11
Lemley, Mark 60–61, 62
Lerner, Joshua 58–9, 62
licensing 66, 200
Lippoldt, Douglas 66
Litman, J. 67

Macao
utility models 33
Machlup, Fritz 42
Malaysia
patents 169–70
utility models 34, 170–72
empirical and policy analysis 174–6
examples of granted claims 173–4
Mandeville, T. 7
market failures 15
mercantilism 13
Mexico 188–9, 192
inventor certificates 189–90
patents 189
utility models 190–91
Mill, John Stuart 14
Millennium Development Goals 5
minor innovations 17
ideal law for minor innovations 199–200
Mowery, David C. 59
Mytelka, L. K. 6

national system of innovation (NSI) 9
Nelson, R. R. 11
Netherlands
economic and innovation climate 23
patents 25, 26

Paris Convention for the Protection of Industrial Property (1883) 20, 49
Park, Walter 66
Patent Cooperation Treaty (PCT) 21, 111, 189
patents 10–11, 13–16
Belgium 25

China 153–6
European Union (EU) 25–6
France 25
Germany 27–8
Indonesia 182
Ireland 25
Japan 42
historical background 142
Japanese and foreign applicants 147
numbers 146
justification 195–7
Malaysia 169–70
Mexico 189
minor innovations and 17
Netherlands 25, 26
Philippines 183
public goods and 15
Singapore 76–82, 100, 102, 117, 186–7
foreign ownership of patents 96–7
ICT and biotechnology patents 93–6
triadic patent families 91–2, 93
Taiwan
historical background 163–4
substantive law 164–6
Thailand 180
United States of America 57–61, 198
Vietnam 176–7
petty patents *see* utility models
Philippines
patents 183
utility models 34, 186
empirical and policy implications 185
substantive law 184
Philips 10
policy considerations 64
copying as creative imitation or misappropriation 66–8
developing countries 197–9
accretion approach 198
emulation approach 198–9
status quo approach 197–8
final options 69
historical and economic climate of a country 69
import/export of IP products 65–6
nature of innovator and innovation 68–9
right type of intellectual property institutional order 64–5
Portugal
utility models 27
public goods
patents and 15
public sector
R&D and 23

regulatory capture 60–61
research and development (R&D)
public sector 23
Singapore 77, 78, 117
reward theory 14
Rio Summit 5
Rosenberg, Nathan 59

Schumpeter, J. A. 6, 8
Sen, Amartya 4
Siemens 28
Singapore 75, 117–18
case studies 75, 110–17
electronic systems company 113–15
healthy lifestyle products company 110–13
water treatment company 115–17
copying 104
copyright 102, 103
innovation 83–8
benchmarking 90–97
international comparisons 88–90
level of innovation 99
methodology of study 73–5
case studies 75
national survey 73–4
targeted survey 74–5
patents 76–82, 100, 102, 117, 186–7
foreign ownership of patents 96–7
ICT and biotechnology patents 93–6
triadic patent families 91–2, 93
research and development (R&D) 77, 78, 117
survey findings 97–110
demographic profile 97–9
level of innovation 99
need for utility model 105–7
statistical testing 108–9
usage of intellectual property system 99–105
trademarks 100, 102

utility models 102, 105–7, 111, 117–18
small and medium enterprises (SMEs)
utility model and 38, 68
Smith, Adam 13
social interaction
innovation and 8
Spain
economic and innovation climate 22
utility models 26, 27
sui generis laws 55–6
sustainable development 4–5
Sweden 24
economic and innovation climate 22, 23

Taiwan 152
patents
historical background 163–4
substantive law 164–6
utility models 33, 162–5, 186
empirical and policy analysis 166–8
substantive law 164–6
technology transfer 66
Tesfachew, T. 6
Texas Instruments 10
Thailand
design law 180
utility models 34
empirical and policy implications 181–2
substantive law 180–81
Thomas, Jay R. 62
trademarks
Singapore 100, 102
triadic patent families 91–2, 93
TRIPS Agreement (1994) 21, 154, 189, 198
Turkey
utility models 33

uncertainty 40–41
unfairness 40–41
unfair competition 43, 49–55
interactions with other laws 56–7
United Kingdom
design law 24, 43, 47, 48, 55, 198
intellectual property institutional order 65
sui generis laws 55
utility models 65
United States of America
design law 44–5
economic and innovation climate 22, 23
federal trade dress law 50–52
need for second tier system 61–3
patents 57–61, 198
sui generis laws 55
unfair competition 50–52, 62
utility models 18–20, 43
Australia (petty/innovation patents) 18, 32, 119, 126, 139–40, 141
empirical analysis 127–8, 131, 134–8
features 122–3, 125–6
history 120–21
objectives 121–2, 125
petty patents 119, 126, 139–40
review 123–5
Cambodia 34, 185–6
China 32, 153, 154–6
empirical analysis 156–62
policy implications 162–3
copying as creative imitation or misappropriation 66–8
costs and benefits 39–42
benefits of copying 41
psychology and bargaining tools 42
uncertainty and unfairness 40–41
examples 20–35
Africa 31, 35
Asia/Pacific 31, 32–4
Europe 22–7
international conventions and multilateral agreements 20–21
ideal law for minor innovations 199–200
Indonesia 34, 182–3
innovation culture and 150–51
interactions with other laws 56–7
Japan 32
correlation between UM and innovation culture 150–51
decrease of UM registrations 145–8
historical background 142–3
individual *versus* company applicants 147
Japanese and foreign applicants 148
procedure and examination 143–5

scope of protection 145
subject matter and criteria of protection 143
Korea (South) 32, 148–50
correlation between UM and innovation culture 150–51
Malaysia 34, 170–72
empirical and policy analysis 174–6
examples of granted claims 173–4
Mexico 190–91
Philippines 34, 186
empirical and policy implications 185
substantive law 184
policy options for developing countries 197–9
accretion approach 198
emulation approach 198–9
status quo approach 197–8
Singapore 102, 105–7, 111, 117–18
Taiwan 33, 162–5, 186
empirical and policy analysis 166–8
substantive law 164–6
Thailand 34
empirical and policy implications 181–2
substantive law 180–81
theory and practice 35–9
Vietnam 34, 177, 186
empirical and policy implications 178–80
substantive law 177–8
Uzbekistan
utility models 33

Vietnam
patents 176–7
utility models 34, 177, 186
empirical and policy implications 178–80
substantive law 177–8